"Mel Bost recognizes that many activities that people do can be more successful through using project discipline. In this book, he describes the elements necessary to successfully build and manage a project team. He illustrates well through real life examples from many fields to illuminate the core principles of project management success."

—**Todd Allen**, *Chair*, Department of Nuclear Engineering and Radiological Sciences, The University of Michigan, Ann Arbor

"High-performing PMOs start with the Ends to be Achieved (outcomes) and then develop the Means to the Ends (outputs). Poor-performing PMOs rush into the Means to the Ends and quickly snatch PMO defeat from the jaws of victory. Congrats to Mel Bost for providing a foundational approach to help organizations start and maintain high-performing PMOs."

—**Mark Price Perry**, *PMO Manager*, Uvoider

"I read this book after teaching a course that covers project management at New York University, and it helped me see this book's value. Many of us, particularly in small- or medium-sized businesses and not-for-profits, think we know how to get things done. Unfortunately, the methods that work individually lead to wasted time and poor results. Mel Bost borrows from a wide range of sources and puts them all into guidelines that you can follow. Designing projects is a skill. It requires attention to learn. It starts when you ask: 'What are we trying to achieve?' With this book, you can aim higher."

—**Art Kleiner**, *co-author*, The AI Dilemma: 7 Principles for Responsible Technology and author, The Age of Heretics and Who Really Matters

Understanding Project Practices and Processes

Understanding Project Practices and Processes distills the author's experiences in developing program management offices (PMOs) and an IT project office to guide project managers who may be facing challenges in delivering project goals and managing teams and stakeholders. With insights into the behavior of project teams, project managers, stakeholders, and project organizations, it brings to life such processes and practices of project management as risk management, innovation, and design thinking.

In presenting principles dedicated to achieving successfully good, sound business and technological maturity, the book explains the following:

- Using lessons learned to improve processes.
- When to implement process improvement to better attain goals.
- Process capability maturity as a continuous process of improving and maturing a process to achieve consistent and repeatable results.
- Distinction between processes and projects.
- How projects turn a strategy into action.
- Key roles of communications and technology.

Research and real-world scenarios explain the practice of project management. With in-depth coverage of project management offices, knowledge management, change management, and project best practices, this book is an indispensable guide for new, as well as veteran, project managers.

Mel Bost is a project management consultant specializing in project closeout and lessons learned, as well as process improvement, best practices, and benchmarking. He has taught "Project Management for Research" to post-graduate students at Arizona State University, as well as developed new approaches to the research process.

He was formerly a practice leader in BOT International's PMO Practice and is a specialist in PMO best practices, Project Lessons Learned, and program management. He is experienced in all aspects of project and program management, including strategic planning, design thinking, knowledge management, risk management, and business process analysis. Bost has successfully developed the processes, standards, procedures, and organizational structures for the PMOs of several major corporations.

Before becoming a consultant, Bost worked for a number of large national and international companies, including Exterran Corporation, ConocoPhillips, Phillips Petroleum, Tosco Corporation, UNOCAL, Atlantic Richfield Company, and Ford Motor Company. Bost began his career in the industry with Ford Motor Company, where he directed product programs for the North American and European markets. He was also instrumental in a group that introduced finite element analysis in Ford Product Engineering using NASA programs.

Understanding Project Practices and Processes

Mel Bost

CRC Press
Taylor & Francis Group
Boca Raton London New York

CRC Press is an imprint of the
Taylor & Francis Group, an **informa** business

First edition published 2025
by CRC Press
2385 NW Executive Center Drive, Suite 320, Boca Raton FL 33431

and by CRC Press
4 Park Square, Milton Park, Abingdon, Oxon, OX14 4RN

CRC Press is an imprint of Taylor & Francis Group, LLC

Library of Congress Cataloging-in-Publication Data
Names: Bost, Mel, author.
Title: Understanding project practices and processes / Mel Bost.
Description: First edition. | Boca Raton, FL : CRC Press, 2025. |
Includes index. | Summary: "Understanding Project Practices and Processes distills the author's experiences in developing PMOs and as an IT Project Office to guide other project managers who may be facing challenges in delivering project goals and managing teams and stakeholders. With insights into the behavior of project teams, project managers, stakeholders, and project organizations, it brings to life such processes and practices of project management as risk management, innovation, and design thinking"– Provided by publisher.
Identifiers: LCCN 2024047974 (print) | LCCN 2024047975 (ebook) |
ISBN 9781032973579 (hardback) | ISBN 9781032974224 (paperback) |
ISBN 9781003593645 (ebook)
Subjects: LCSH: Project management.
Classification: LCC HD69.P75 .B673 2025 (print) | LCC HD69.P75 (ebook) |
DDC 658.4/04–dc23/eng/20241011
LC record available at https://lccn.loc.gov/2024047974
LC ebook record available at https://lccn.loc.gov/2024047975

ISBN: 9781032973579 (hbk)
ISBN: 9781032974224 (pbk)
ISBN: 9781003593645 (ebk)

DOI: 10.1201/9781003593645

Typeset in Minion
by Newgen Publishing UK

To my granddaughters,
Quinn Penelope Seaton
and
Jordan Persephone Seaton

Contents

Preface

Project management has played a major role in my life for more than fifty years. It has taken me to Great Britain, Panama, the United Arab Emirates, and to most of the United States. I have participated in merger teams to develop program management office (PMO) organizations for several major corporations. The evolution of the PMO as a best practice organizational structure has placed the responsibility for project manager and team development and performance squarely on the shoulders of the PMO. This has added new dimensions to the development of competencies and capabilities for project managers and teams.

This book is a collection of essays written since 2010, which summarizes my experiences, insights, and interpretations of the evolution of the project management discipline over that fifty-year period. Many of you have followed my blog "Mel Bost PMO Expert" over the years. This book is a collection of and addition to those essays.

This volume includes new project research topics that have added to the richness of these essays. Every essay informs and lays the groundwork for the reader to delve deeper into the research.

Readers will also note a number of references to Louis Tice who was the president of the Pacific Institute in Seattle for many years. I was privileged to be submerged in his class in 1982 at Atlantic Richfield Company (ARCO) entitled "Achieving Your Potential" or "Investment in Excellence" and that experience laid the groundwork for my beliefs and values regarding the project manager experience. Louis Tice's work has guided me for over forty years.

In addition, the work I have done in recent years with the Greater Phoenix SCORE (Service Corps of Retired Executives) in mentoring small business has guided my learning and research. In focusing on the role of projects and business owner initiative in small business, I quickly realized how similar the experiences of small business teams were to those of project teams in terms of leadership, performance, competencies, and outcomes.

Project management as a business discipline is usually conducted in large, complex organizations. Highly successful companies such as Accenture have studied what contributes most to success and have actually designed road maps to guide their progress. Three topics have risen to the top of Accenture's road maps: (1) leadership, (2) programs and policies, and (3) environment. In this book, I have devoted many essays to each of these topics. My hope is that the reader will keep these in mind as a road map to guide in implementing these ideas.

With regard to my project experiences, I have seen new competencies derived from the influences of the cognitive science and psychology fields, the rise of the PMO as a best practice organizational structure in project-based organizations, the application of computer technology to add a new dimension to the resource/schedule/planning/performance dynamic, the development of collaboration and simulation software, and the feedback of project lessons learned to facilitate process improvement These innovations have rocketed project management to new heights as a management tool in every facet of business. Readers will also note the emphasis on "learning" and the familiar expression "stories are the essence of learning." This book is, more than anything, a collection of stories. Absorb them, retell them.

You only have to review the job descriptions in today's modern corporations to see how many project competencies are reflected in today's work environment. Project-based initiatives and outcome-based results are the norm for successful organizations.

Project leadership and performance-based initiatives are now the hallmark in forward thinking organizations. How project managers interact with their teams and all stakeholders is key to success. This collection of essays provides project professionals with a hands-on approach and answers to their most pressing daily issues.

Open this volume and begin reading any essay. It will command your attention while informing and enriching your project experience. You will embrace the research in a uniquely human way. This volume is for you as surely as if I were sitting next to you as you read.

The significance of this book to a reader today lies in the fact that I have presented actionable solutions in this book to the problems, issues, and opportunities facing professional project managers and PMO facilitators. For example, with regard to my essay "What do I Do If My PMO Can't Execute?," I present an analysis using systems thinking principles to identify problem issues and present possible solutions that can be implemented within a PMO. These are real life problems I have identified over my fifty-year project career.

Organizational psychologist Adam Grant has been studying the application of "rethinking" in organizations as described in his best-selling book *Think Again: The Power of Knowing What You Don't Know*. "Rethinking" is a form of lessons learned, as I have described its use in my previous book, published in 2018, and in this book. According to Grant, "'rethinking' is the foundation of a learning organization." A learning organization culture is essential to project management and PMO maturity. Project managers and PMO facilitators thrive in a learning organization.

In essence, project management is decision-making with many data points, many potential decision-makers, and a myriad of options to consider. Many organizations today have moved ahead with incorporating AI techniques into their project methodologies. For these organizations, this book presents a database with many of the decision points and options for action which will feed the project management/AI decision processes. Decision automation using cognitive techniques will mature the organization's project processes rapidly.

Thank you for your attention.

Mel Bost

Cave Creek, AZ

About the Author

Mel Bost is a project management consultant specializing in project closeout and lessons learned, as well as process improvement, best practices, and benchmarking. He has taught "Project Management for Research" to postgraduate students at Arizona State University, as well as developed new approaches to the research process.

He was formerly a practice leader in BOT International's Program Management Office (PMO) Practice and is a specialist in PMO best practices, Project Lessons Learned, and program management. He is experienced in all aspects of project and program management, including strategic planning, design thinking, knowledge management, risk management, and business process analysis. Bost has successfully developed the processes, standards, procedures, and organizational structures for the PMOs of several major corporations.

Before becoming a consultant, Bost worked for a number of large national and international companies, including Exterran Corporation, ConocoPhillips, Phillips Petroleum, Tosco Corporation, UNOCAL, Atlantic Richfield Company, and Ford Motor Company. Bost began his career in the industry with Ford Motor Company, where he directed product programs for the North American and European markets. He was also instrumental in a group that introduced finite element analysis in Ford Product Engineering using NASA programs.

With Atlantic Richfield Company, he directed Fuels Research and Development Projects. With UNOCAL and Tosco, he led project and PMO programs. He was part of a global merger-transition team that developed business analysis and best practices for a new PMO before the merger of Conoco and Phillips Petroleum. He also led a global team to create a blueprint for implementing a PMO for Duke Energy Field Services and LUKOIL.

Bost has worked with the PMO Executive Council in Washington, DC, to define best practices for PMOs and with the Advisory Board of the University of Arkansas, Walton School of Business, Information Technology Research Institute. He is the author of the highly regarded blog, "Mel Bost PMO Expert," which addresses the structure, activities, and behavior of a PMO environment, and has also recorded numerous podcasts with Wayne Thompson of "Project Management War Stories."

In 2011, Bost was invited by the Panama Canal Authority to conduct two training courses in Panama with their Construction Division project managers, engineers, and other project personnel on the topic "Project Closeout and Lessons Learned." The Panama Canal Authority was engaged in a $5 billion program to expand the Canal, which was completed in 2015.

In 2012, in a cooperative effort between BOT International and Collaboration Management and Control Solutions headquartered in Dubai, Bost conducted a week-long training course in Dubai on Project Lessons Learned for several companies from Morocco, Saudi Arabia, and the United Arab Emirates. Collaboration Management and Control Solutions is one of the largest providers of project management training in the Middle East and Far East.

In 2012, in a joint effort by BOT International and ProjectManagement.com, Bost conducted one-day Project Lessons Learned training sessions in several US cities. ProjectManagement.com is one of the largest online networks of project managers in the world.

Bost completed his BS in Physics and MS in Engineering from Georgia Tech and his MS in Engineering Science and MBA degree from the University of Michigan, Ann Arbor. He lives with his wife in the Scottsdale, Arizona area, and is an active volunteer with the University of Michigan Club of Phoenix.

Acknowledgments

This book is the product of being immersed in the project management discipline for more than fifty years. However, its roots were planted when I was still in high school. I loved writing and sports, so many Friday evenings were spent developing an imaginary article summarizing the events of the local high school football game in the same format I often read in the Sunday Sports Section of the local newspaper. Then, as a junior in high school, I was Sports Editor of the school newspaper. This experience allowed me to develop a style and comfortable expressions. I knew very early that I wanted to express myself with the written word.

My daughter Christina Bost Seaton deserves credit for reading and editing every word and making intelligible many phrases and paragraphs.

I would like to acknowledge the contributions of two of my mentors, Professor Ziya Akcasu and President Emeritus and University Professor of Science and Engineering James Duderstadt. They were both Professors at University of Michigan who influenced me greatly and who, across the decades and long after my graduation, continued to guide and advise me, and inspire me to approach life with the curiosity of an academic. They both instilled in me a love of research and the study of physical systems, which exhibit the influence of feedback to produce the final process outputs. The study of physical systems later led me to look for nonphysical systems, such as business processes, where performance-based feedback in the form of lessons learned could be used for process improvement. My thanks also to Todd Allen, the current Chair of the Department of Nuclear Engineering and Radiological Sciences at the University of Michigan, for his continued support and encouragement of my research and writing.

I would also like to thank my CRC Press Editor John Wyzalek for his encouragement and steadfast loyalty to this book.

The teachings of Louis Tice also influenced my thoughts about the role of cognitive processes on project manager behavior and performance.

Finally, to all my colleagues in the many corporations I have been a part of, thank you for your collaborative efforts and your steadfast friendship.

1

Introduction

Stories are the essence of learning

—Mel Bost

THE PROJECT COMMUNITY

I want to share my experiences developing several Program Management Offices ("PMOs") and an IT Project Office with others who may be facing the same issues and similar situations, so that they may benefit from my insights into the behavior of project teams, project managers, stakeholders, and project organizations.

What do I mean by the word "community" and the phrase "project community"? In my mind, a "community" is a loosely connected group of people who have common interests in a field or discipline. They are all interested in the growth and well-being of that field or discipline. A community need not be all in the same organization or in the same geographical area. It does, however, need some framework for sharing information and collaboration. Such a framework allows the community to benefit from the work of its members, from the input of outsiders whose ideas impinge on the community environment, from academics, and from other researchers.

Social media and social technology facilitate such a community environment. Rich Maltzman, a practitioner and thought leader in this space, first alerted me to the importance of the project community in our discussion about the evolution of PMOs. PMOs, in general, began as more limited entities, focused on a company's IT projects (the "IT PMO") or as Shared Services PMOs (where the PMO serviced only certain business segments of

DOI: 10.1201/9781003593645-1

the company with a particular "mindset"). Over time, PMOs have become far more ubiquitous, and it is not uncommon to see Enterprise PMOs ("EPMO") that service the entire organization, or several PMOs within the same organization. There are still, however, more focused PMOs designed to serve specific targeted project goals and objectives or requirements (such as Smart Grid with some utilities).

Rich Maltzman took the position in his blog, "Scope Crepe," that regardless of how many PMOs there were in an organization, it was important to maintain a sense of "project community" across the PMOs. Communities contribute to the well-being and ongoing growth of the members of the community.

In her book *Open Leadership*, Charlene Li presents an excellent model of engagement that could apply to either an organization or an individual engaging a community. The hierarchy of increasing engagement proceeds as follows:

1. Watching
2. Sharing
3. Commenting
4. Producing
5. Curating

My own involvement with the project community has followed the same model.

First, I watched as the project community embraced the *Project Management Body of Knowledge* ("PMBOK") and the literature surrounding the progress of projects in a corporation from early simple executions to large-scale executions of tremendous strategic importance to the corporation. At this point, the Project Management Institute (PMI) and its chapters engaged with the "project community" via publications, online websites, and periodic meetings throughout the world.

As I progressed through the stages of Sharing and Commenting, I was reacting to the written and spoken words of others in the "project community" who were contributing to the processes, standards, and methods by which good Project Management Practice was being spread throughout the community.

Now, in the stage of Producing, I am contributing to the literature and culture of the "project community" through my blog, "Mel Bost PMO Expert," and my multiple project management publications.

I feel that the Curating stage lies ahead because I am still a "youngster" when it comes to "contributing." Social mediums, such as blogging, have certainly furthered this journey—and, yes, it is a journey of growth and awakening, and finding out how much I don't know about things I should know. We all can keep learning new skills and techniques over the course of our careers.

Being part of a larger community also helps us maintain our own identities. When my manager at Exterran Corporation informed me in October 2009 that my position had been eliminated in a budget cut, I was amazed by my own loss of identity. Like many other people, I largely associated myself, my work life, and my accomplishments with a Corporation.

If you have seen the movie *Up in the Air*, this experience was very much like sitting across the desk from George Clooney and listening to him say, in his monotone voice:

> You are being terminated from the Company and I will need your access badge. Please be prepared to leave the building as soon as possible with your personal belongings in this box. And don't worry. All your questions will be answered if you read what's in this small packet.

Participating in the project community, however, led me to new opportunities and experiences that I would never have had if I stayed in that position.

The social media revolution is an event that we cannot deny. It has made the community a relatively easy thing to obtain—a community far outside our localities. People are relying on other people—through social media—to satisfy their wants and needs: needs they once satisfied by going to established stores, banks, or other institutions. Charlene Li referred to this phenomenon as a "groundswell"—a social trend whereby people use technologies to get the things they need from each other, rather than from traditional institutions such as corporations.

If everyone were to approach their interactions in a similar manner, they could find the "self-direction," "mastery," and "Purpose" that Daniel Pink writes about in *Drive*.

I urge you to consider how you can contribute and give to the "project community." What skills, experience, ideas, innovations, and specifics can you contribute to everyone who wants to grow and nurture the project discipline in the same manner? What project lessons can you share with another project manager who is facing a similar situation or scenario? What project analysis did you perform that led to the successful completion of some very difficult phases of a project that you could share with your "project community?"

If you are watching, sharing, and commenting on the work of others in some fashion, I urge you to consider "Contributing" through your writing, observations, and insights about projects and programs. Your reward in return for new information, insights, and knowledge will be amazing.

A PRIMER FOR PROCESSES AND PROJECTS

1. No process was ever created that could not benefit from process improvement using lessons learned.
2. A process is a set of activities or tasks which, when performed in a specific sequence, yields a desired result. The most important word in this definition is "desired." If a desired result is not achieved, process improvement is required.
3. Process Capability Maturity is a continuous process of improving and maturing a process so that it achieves consistent and repeatable results.
4. All projects are processes, but not all processes are projects.
5. Projects turn strategy into action for an individual or an organization.
6. The most important competency in project management is communication.
7. Technology is the key to success as a project manager.

These are principles to live by if you are dedicated to achieving good, sound business and technological maturity. These are truly "success principles."

2

PMO Formation, Structure, and Methodology

WHAT IS A PMO METHODOLOGY?

> Either write something worth reading or do something worth writing.
>
> Benjamin Franklin

In its simplest form, a PMO Methodology is a guide that instructs an organization how its project management practices will be conducted in the context of its organizational strategy, structure, and mission. This guide contains standards, processes, procedures, templates, quick reference materials, and an overall structure which facilitate ease of implementation and execution. Suppose we define a project with four phases:

1. Initiation
2. Planning
3. Execution
4. Project Closeout

A PMO Methodology adds a structure which tells the organization what steps to take to complete the project action. In addition, as you proceed through the phases, each topic from a phase is continued in successive phases so that a complete project is the result.

Here is an example of a PMO Methodology:

1. Initiation
   a. Project Charter
2. Planning
   a. Business and Technical Requirements

DOI: 10.1201/9781003593645-2

   b. Risk Management
   c. Communications
   d. Change Management
3. Execution
   a. Schedule
   b. Resources
   c. Dependencies
4. Project Closeout
   a. Project Closeout and Lessons Learned
   b. Knowledge Management

Organizations may add additional structure for unique situations in the context of their businesses and industries.

Methodology is such an important aspect of PMO structure that it is continually evaluated for completeness of coverage of real-world projects.

WHAT ARE THE ESSENTIAL ELEMENTS OF A PROJECT CHARTER?

The smooth and efficient launch of a new project is an event of great importance to an organization or an individual. The Project Charter that goes with the launch is perhaps the most important document in the project methodology.

Essential features of a Project Charter include the following elements:

1. Assumptions
2. Stakeholders
3. Communications
4. Funding
5. Dependencies
6. Impact
7. Lessons Learned

The Assumption section is extremely important because it details the initial inputs to the project before problem-solving, change orders, or decision-making about project direction. Assumptions should be revisited at intervals in the project. It has been my experience over the years that the Assumptions section often goes untouched after Project Charter approval.

Stakeholders should be clearly identified with any explanations necessary so everyone agrees on stakeholder status.

Communications should clearly state who will be the recipient of communication and in what form, that is, report, presentation, status update, spreadsheet, and so on.

Funding will specify what budget and organization is responsible for the costs of the project.

Dependencies should state what deliverables will be made to the project and what outputs can be expected for delivery to other projects.

Impact should state the significance of the project to the organization as well as to the portfolio of other projects active during the same time period.

Lessons Learned from other recent projects will often aid the current project team to incorporate the process improvements into the current project, thus enhancing the overall process performance to meet stakeholders' needs.

The Project Charter should be a standalone document signed off by management. It should be a "referral" document and not a "shelf" document.

SO HOW DO I START A PMO?

Suppose you are a member of a successful IT or other business functional group whose members have successfully delivered value to their larger organizations/corporations through successful project implementation. Inevitably, this success brings additional requirements from the corporation to deliver more value through more projects over time. You begin to feel the pressure because resources may not be readily available or accessible in the time frame you are asked to deliver, teams may not be prepared, and, in general, although you saw it coming, your best efforts just to deliver the last increment of value was enough to keep you occupied.

So what's a body to do?

This is exactly what every organization that decides to ramp up to a Program Management Office (PMO) faces. But the question remains "How do I define just what I need in the time frame I have been given to be successful?"

My answer, which I have leveraged in several situations before, is to take a blank sheet of paper and write a "Funding Authorization Statement" for a single project just as if you were seeking funding today.

An example:

> This funding authorization seeks **$37.5 million** (capital and expense) from the **IT Executive Board** to deliver the **IT Marketing Apps Development Project** for new **Stores** by **December 2026**. The **Business Case** is defined as follows. … The **economics** and **payback** are defined as follows. … The **risks** to the project completion are defined as follows. … The proposed **project team** will consist of the following from **IT** and/or **functional business units**. The **external consultants** proposed for this project are … **Authorization** of this project is part of a larger **portfolio** of projects which have been addressed by the **IT Executive Board** in its meeting of **August 21, 2024**.

Now, such a simple statement defines many things. It defines what project, what funding group, what business case, what risks, and so forth.

Every detail must be provided by resources that reside in the PMO or functional business units interfacing with the PMO.

When the PMO has a number of these Authorization Statements, it can begin to decide if a methodology of handling the portfolio of projects, and the life cycle of the project management process needs revamping to facilitate consistency in the way all the projects are evaluated and authorized.

Every **Bolded Word** (Information) in this Authorization Statement needs to be addressed, including the following:

1. Definition of the **Bolded Word** (Information).
2. Location of the **Bolded Word** (Information), that is, inside the PMO, supporting corporate group and functional business group. This detail will provide the PMO Author with decision points for various pieces of information.

   For example, the **Business Case** will contain financial and economic analyses. Whichever person or department prepares that financial and/or economic analysis sets up a decision point for the PMO Author.
3. At this point, I recommend that the PMO Author actually draw lines from the **Bolded Word** (Information) named in the Authorization Statement to a white space in the margin where the PMO Author will begin to fill in the details of what, where, when, and how for each piece of information.
4. It is important that every **Bolded Word** (Information) in the Authorization Statement be addressed by the PMO Author at this point so that the next stage of maturation in the process will flow smoothly. Those of you familiar with David Allen's *Getting Things Done* will recognize this as the next action.

5. Once all the information and locations are defined for one project's Authorization Statement, then additional Authorization Statements can be added for additional projects until a portfolio of projects is included. At this point, an information piece needs to be added to describe how the portfolio will be handled.
6. Obviously each **Bolded Word** (Information) will include human resources as well as physical and financial resource materials so that a Budget of human resources and physical and financial resources can be developed.
7. Timing will also be a consideration after all **Bolded Words** (Information), and Resources have been defined. Now the "process" is beginning to look like a "Funding Authorization Process" and a flow of information to support the "Process."

What does your "clean sheet of paper" look like? Have you given it some thought lately? If you are an organization looking to move to that next level of superior project service and "value addition" to your corporation, think about that "blank sheet of paper." Your business context will determine what elements need a place in the PMO and what must come from the business functional areas and perhaps even what must come from external stakeholders.

ADDITIONAL THOUGHTS ON HOW TO BUILD A PMO FROM THE "GRASSROOTS LEVEL" USING A CLEAN SHEET OF PAPER

I collaborated on a podcast with Wayne Thompson, the host of the very popular blog "Project Management War Stories," and we decided to revisit the topic of building a PMO from the "grassroots level."

Here are some additional thoughts, based on our discussion, that you should keep in mind when using the "clean sheet of paper" approach:

1. The importance of "commitment" vs. "compliance" in the design and success of a PMO.
2. Issues of "accountability" vs. "authority" in PMO organizations.
3. Designing a PMO for a small entrepreneurial organization vs. designing a PMO for a larger, more disciplined, bureaucratic organization.
4. The importance of including all Stakeholder voices in the design and success of a PMO.

5. Building capability as a critical success factor in the design of a PMO. What other areas would you add specific to your experience?

BUILDING FUTURE PMO TECHNOLOGY DEVELOPMENT CAPABILITY IS THE KEY TO ADDING VALUE

As I work with project managers in many fields, it is becoming more apparent every day that, in the future, new technology development will play a greater role in delivering value from project management. Of course, new technology development has been a major part of many projects. Witness the development of electrical components and control systems in the Panama Canal project from the early 20th century, as I discussed in my Lessons Learned book. Today, we only have to look at the technology and new business fields and sources such as CNBC TV and CNN to hear about new "apps" and new technologies that are adding new, or replacing old, functionality in products and services.

Look no further than Wal-Mart's growth into e-commerce and Amazon's growth into more brick-and-mortar solutions to find that new technologies are at the heart of business and consumer demands.

So, companies are struggling with the question "Do I grow this new technology by acquisition or by internal development of the skills, talents, and structure to compete in the future?" The answer to this question lies in another question:

"How do I develop the 'capability' that will enable me to deliver my desired end value?"

This was much the same question many years ago when organizations began to concentrate their project management skills, practices, and methodologies into an internal group that became known as the "Program Management Office (PMO)." The PMO quickly became a standard and "best practice" for delivering consistency and repeatability of project results. The PMO is often referred to as a "capability-based" organization because it was constructed to DELIVER projects successfully.

CAPABILITY is the combination of people (resources, talents, leadership); processes; technologies; and organization (structure, networks), which allows an individual or organization to deliver its intended objectives. The blueprint covers all those components, but not separately. It covers how they will fit together. There is also an accompanying plan that specifies the people who will build pieces of the capability, the targets and incentives that govern their actions, and a timetable for the implementation.

Recent studies by Booz and Company consultants have focused on "capability building" as the key to closing what they refer to as the "strategy-to-execution" gap that has plagued companies for many years. Many times in the past, a great strategy never seemed to materialize into actionable results. This was termed the "strategy-to-execution gap." Their analysis has been documented in the 2016 book *Strategy That Works: How Winning Companies Close the Strategy-to-Execution Gap* by Paul Leinwand and Cesare Mainardi, with Art Kleiner. Their process for building new capability focuses on answering the following questions:

1. What is the capability?
2. Why is it valuable?
3. How would it be different from what we have today?
4. Describe a day in the life of this capability; what does it look like?
5. What is required to make it work?
6. For the "capabilities system," what does the business case look like?
7. How does this capability fit with others in the capability system?

The focus of this analysis was to build meaningful and lasting capabilities and to foster cross-functional relationships between corporate competencies that complement each other. In short, their theme was "build capabilities before seeking results."

To summarize, PMOs in the future will need to develop new technology capabilities systems which will add to their project delivery capabilities. Of course, specialized PMOs in companies such as biopharmaceuticals or information systems will need to consider industry-specific technologies to insure their "table-stakes" status. This new reality may require a "maturity curve" approach in implementation to be successful in the long term.

WHY THE SUDDEN PROLIFERATION OF PMO ORGANIZATIONAL STRUCTURES IN PROJECT-FOCUSED ORGANIZATIONS?

I have observed that in the last several years project-focused organizations have increasingly expanded the use of (or created) PMOs within their organizations.

Historically, the PMO concept first appeared in Information Technology or Information Services organizations, principally because those structures advocated and sponsored more disciplined and consistent/repeatable processes including change management, applications development, and project management processes. From IT and IS organizations, the PMO then became more widespread in Shared Services organizations which served the larger corporate environment by handling common functions such as Procurement, Financial Services, Facilities Management, and so on. Next, the Enterprise PMOs (EPMOs) appeared to serve the entire business organization or some major business/functional portion of a very large organization.

Most recently I have observed that:

1. Both Home Depot and Lowe's, two major Fortune 100 home building supply organizations, have formed IT PMOs.
2. The Federal Reserve Bank of New York has expanded its IT organization to handle projects through a new IT PMO structure.
3. Progress Energy, the electric utility serving portions of North Carolina and Florida, has recently introduced a Smart Grid Program Management Office (SG PMO) to handle all projects related to its Smart Grid business.
4. Apple has introduced a PMO in its financial operations, reporting to the CFO.
5. Cisco, ADP, Sirius/XM, Microsoft, Time Warner Cable, BP, and many other major firms have employed variations of PMOs to serve portions of their businesses.

The list goes on and on.

It would be very easy to say that this proliferation of PMOs is simply due to organizations maturing to recognize the value added by the PMO structure to the organization. However, my background in science and engineering led me to search for the underlying issues and ask several knowledgeable people who have studied PMOs from within and without for years.

I spoke with Rich Maltzman (who writes a great blog about project management entitled "Scope Crepe" and who also works in a PMO of a major telecom company) and asked him about his impressions of PMO proliferation. Likewise, I asked Wayne Thompson (author of the well-known Project Management War Stories blog) for input on this subject. Over the last several years, Wayne has interviewed a number of successful PMO managers for a series of podcasts about the functioning of their PMOs in small to large corporate settings.

Rich Maltzman said:

> From my Scope Crepe and consulting work, I see that PMOs have long been the bastion—not only of IT but also of enterprise-level project organizations. I think, in fact, there's a danger in not trying to build a community that goes beyond the IT or customer-facing PMs in an enterprise. The overriding PM discipline and the community of PMs in an enterprise should be respected as a community and there should—even MUST—be sharing between the IT PMs and those who run the enterprises' projects with customers (i.e., deploying a network, introducing a new pharma product). So, a PMO at the enterprise level has value in that it builds a strong sense of community for PMs, honors the discipline, shares artifacts, and allows for the career growth of project managers—including job paths that may intertwine between IT and other enterprise functions.

Wayne Thompson remarked:

> PMOs originated in IT and since IT was traditionally looked upon as a "cost center," the same view of that role carried over to the IT PMO. But clearly, moving to the enterprise level with the PMO, organizations have recognized that the real takeaway in employing a PMO at that level is that it clearly adds value (rather than merely incur cost) to the enterprise's strategies and should be evaluated as such. Enterprise PMOs deliver concrete results in a disciplined manner. In my blog work, I have found that PMOs have been deployed for any number of reasons depending on the business and industry context for that company. And their scope is still expanding.

This same reasoning applies to organizations just entering the project management discipline area as well. A recent article "Dechert Puts Every Project Manager Through Project Management Training" (from *The Legal Intelligencer*) outlined the fact that law firms are now embracing project management as well. Traditionally, many law firms felt that they were different and that the work they did was somewhat unique and could not be managed using a discipline such as project management. Pamela Woldow, a lawyer with the legal management consultant firm Altman Weil, gathered feedback from general counsels and in-house counsels that they wanted their law firms to operate more like businesses and deliver their services more efficiently and cost-effectively. Consequently, Pamela has been developing and delivering project management training to partners as well as to associates.

More and more law firms are seeking training in basic project management principles, and Pamela believes this trend will continue as legal firms seek to satisfy the needs of their clients.

Once again, value addition from a consistent, repeatable process is the key—just as it has been in the ramp up of PMO structures in major project-focused firms.

So what does this all mean?

In William Bridges' groundbreaking 1994 book *Job Shift: How to Prosper in a Workplace Without Jobs*, in which he summarizes the evolution and transition taking place in the traditional notions of work, jobs, and careers, Bridges states that:

> The single organization pattern that is free from this built-in bias (toward maintaining the status quo in organizations) is the project cluster. Project oriented structures offer the important advantages of tailor-made design to fit unique tasks, flexible resource commitments, defined termination points, and an absence of enduring commitment that encourages a resistance to innovation.

In the years since this statement was made, I believe that these "project clusters" have taken on the new structure of PMOs and that PMOs have been deployed in organizations wherever a "cluster" of projects has been the norm. The expectation is that PMOs will add lasting and strategic value to the organization.

We are going to see more specialized PMOs in organizational settings as time goes on.

I would welcome some comments on these observations as well as supporting or alternative reasoning for the sudden proliferation of PMO structures in organizations. Your comments are welcome.

HOW TO ESTABLISH AN INNOVATION PMO

It has been difficult to pick up any literature on business strategy and business growth recently without seeing the word "innovation" splattered all over the headlines and content. Innovation is such a catchy word these days because the idea inspires people; it connotes taking the innovator to a new height of success. Likewise, there are some companies whose names seem to drip with the word "innovation" whenever you encounter the company's name: DuPont, Procter & Gamble, and Apple, for instance.

A.G. Lafley, in his book *The Game Changer: How You Can Drive Revenue and Profit Growth with Innovation*, defines innovation as "the process of converting or turning new ideas into revenue and profit." Similarly, many authors in the PMO field have defined the PMO as a distinct organizational structure that can be used to drive innovation through project management and product development.

An article in the Winter 2010 issue of ***Strategy + Business*** magazine, entitled "The Global Innovation 1000: How the Top Innovators Keep Winning" (www.strategy-business.com/article/10408, accessed October 2, 2024), was based on an ongoing Booz & Co. study of innovation in the most successful companies in the innovation game today. The goal of the study was

> to examine the capabilities needed to maximize the impact of the company's innovation efforts in good times and bad, and to highlight the benefits both of focusing on the short list of capabilities that generate differential advantage and of clearly linking the specific decisions within innovations to the company's overall capabilities system and strategy.

Their premise was that "innovation capabilities enable companies to perform specific functions at all the stages of the R&D value chain–ideation, project selection, product development, and commercialization." The consultants asked the respondents in the Global Innovation 1000 survey which capabilities were most important in achieving and sustaining success in innovation.

Surprisingly, the study found that all the successful companies surveyed depended on a *common set of critical innovation capabilities*. These include the ability to gain insight into customer needs and to understand the potential relevance of emerging technologies at the ideation stage, to engage actively with customers to prove the validity of concepts during product development, and to work with pilot users to roll out products carefully during commercialization.

The study's authors also found important the ongoing assessment of market potential during the project selection phase.

> I have discussed the rise of PMOs for specific purposes in organizations that recognize the special value-added by having an organization focused on project management capabilities as a means of converting strategy into action. For example, some utilities have formed Smart Grid PMOs to handle Smart

Grid projects. In an organization that is determined to succeed and grow in its industry, with best-in-class products, services, or both, why not consider creating an entity called the Innovation PMO? What would the characteristics of such an entity be?

First, since feedback and research from consumers, users, and other stakeholders are critical to understanding "what to innovate for," the Innovation PMO would establish its own unique source of feedback research within the context of its own industry or consumer setting. This is a key element of success. Unfortunately, very few PMOs are currently doing a good job in this area, but to become an Innovation PMO, they must.

Second, the Innovation PMO has leveraged its key supplier and vendor relationships. It is known that frequently, the unsolicited feedback provided by suppliers and vendors provides a fresh look as to where the market is headed, especially in innovation scenarios.

Third, the Innovation PMO has tailored the critical innovation capabilities discussed in the Booz & Co. study to the internal business context of the organization. This tailoring process is analogous to the benchmarking process, which is used by leading PMO organizations, such as American Express and Procter & Gamble, to establish best practices. Like best practices, to be successful, innovation capabilities can't be lifted verbatim without being tailored to the organization's business context.

Margo Visitacion of Forrester has also written about the expansion of the PMO concept to innovation in organizations in her recent Forrester paper "Involve Your PMO to Find the Right Match for Innovation Opportunities." I would recommend you read her analysis as well.

Innovation is certain to be a topic embraced by more organizations looking for successful growth-promoting projects in their industries. Your role as a PMO practitioner is to find a spot where you can contribute to that success. Remember, "Change Creates Opportunity." Now, go find that opportunity.

HOW DO YOU GAUGE THE MATURITY OF YOUR PMO ORGANIZATION?

Last week another PMO practitioner in a related industry asked me the question "How do you know how mature your PMO is at any given time?

Is there a measure or gauge for maturity that you can use to determine your current state?"

This is an extremely timely question in today's project and program management environment. And I am sure there are many different answers to this question that might address some or most aspects of maturity.

Before we tackle this question, however, let's ask another question: "Why does it matter?"

Those of you who work in software and hardware development organizations that supply the government with products and services know that the standards established by the Software Engineering Institute's Capability Maturity Model (SEI CMM) are used by the government procurement functions to gauge the maturity of the project management processes in those supplier firms. So, in that context, maturity is relevant, adds value, and has a recognized measurement standard.

To those organizations that are not suppliers to the government, and therefore not subject to the application of SEI CMM standards, a slightly different measurement process is applicable. In some cases, your PMO organizations may have chosen to use the SEI CMM framework as a "guideline" for gauging maturity. And, if so, it is really up to the individual PMO to decide whether to implement plans to move up the maturity ladder a certain distance. Indeed, it is really up to each individual PMO to decide whether it wants to move all the way to Level Five. Business context and cost-benefit for improvements should be the gauge here.

Several years ago, PMI launched a survey study addressing the question "Is Project Management relevant or a Worthwhile Discipline for organizations to invest in along with their other business processes?" After a great deal of study and review, the answer was that project management is worthwhile as long as it adds value to the business context in which it is applied. This issue of value is key when determining maturity, and your organization's pace of development as well.

There are many ways to address maturity. The Software Engineering Institute, working with PMI, has developed an Organization Maturity Model (OMM) which uses the five-stage SEI CMM model as a framework, and Categories of metrics such as People, PPM Processes, Technology, Financial Management, and Relationships. Many companies have chosen to define a plan for improving their maturity level using the typical behaviors and characteristics in the matrix formed by the SEI CMM five stages and the Categories above. This is an entirely acceptable method to use.

I would propose a slightly different method which can be tailored more to your organizational context and to your interpretation of activities and behaviors within your own project management processes.

Suppose, for example, that we define a very basic project management process as having four phases: Initiation, Planning, Execution, and Close. Compare your PMO organization's PM process to this very basic model and make sure that it includes each phase. Obviously, if you have a PMO in place, your organization has advanced its project processes to a point where there is probably a lot more detail than just these four phases.

Now, give your assessment of the value provided by your current PMO process and gauge it by the value provided by the four-step or phase process given above. Think about how the performance of the process is measured and the various portions of PMBOK that are included in your process. Ask yourself the question "Is there any one area of PMBOK that is not providing the expected value?" If there is more than one area, then prioritize the work, giving the highest priority to a plan to improve the PMBOK area most in need of upgrading.

Recall that the PMBOK Management areas are:

Integration
Time (Performance vs. Schedule)
Cost (Performance vs. Budget)
Quality
Human Resource
Communications
Risk
Procurement (including materials and vendor services).

For example, Project Controls usually encompass PMBOK Management areas such as Time, Cost, and Quality. If the Project Controls aspect of the PMO project performance is not meeting expectations, then changes to the Project Controls area can be made to upgrade the maturity. If the vendor services portion of Procurement is lacking, it can be addressed with initiatives focusing on the Vendor Management Process. Signs of the need for upgrade often appear in Audits and Control Reviews of projects. Stakeholder surveys are also another good source of feedback concerning what is going well and not so well with project execution, planning, and delivery.

This type of maturity assessment and planning is more actionable than one derived from using the Organizational Maturity Model cited above. Usually, an organization will use benchmarking to assess its processes versus other recognized or accepted processes from other authoritative project references. Also, organizations such as Gartner, Forrester, or the Corporate Executive

Board are excellent resources for determining best practices which can be applied to the business context of a specific PMO.

If you apply this technique in your PMO, you will be adding more "depth" or layers of project management "processes" to your PMO PM process. This "depth" can be added until you feel that the full value of the PM process has been reached. From a continuous improvement perspective, every organization should continually be asking itself the question "What do I need to address to improve overall PM value to the organization and to the stakeholders?"

Where does this type of maturity assessment and plan come from? It comes from my observations of a number of PMOs at various stages of maturity, each trying a different process of assessing its maturity and then moving forward with appropriate plans. What I have done here is synthesize a process whereby easy steps can be taken by most PMO members to move the PMO forward in maturity.

"NEXT PRACTICES" INITIATIVES AND PMO FORMATION

The usual way that PMO impact the firm's strategy is through the execution of their value proposition. The execution of projects yields tangible, concrete assets; new processes; or principles of operation that guide the firm and add long-term value to the shareholders. How can PMOs more effectively deliver long-term value and create a competitive advantage for the company?

We have considered several component issues of PMOs in this chapter, such as, for example:

- "Best practices" in a project context
- Evolution of PMOs from a "cost center" IT PMO perspective to EPMOs to specialized PMOs to satisfy specific special business or functional needs in an organization
- "Reframing" project scenarios to ensure we are tackling the right problem with a project
- Design of PMO structures and functions using a "clean sheet of paper" approach

To further explore the boundaries of PMO theories, let's leverage the insights of Professor C.K. Prahalad, a management and strategy guru who was recognized worldwide for his strategic intent, core competency, and "bottom of the pyramid" theories. He was a university professor in the Ross School of Business at the University of Michigan (which was my MBA program).

In an April 2010 article for Harvard School of Business entitled "Best Practices Get You Only So Far," Prahalad discussed "innovation" in corporations from the perspective of looking beyond "best practices" to find what he termed "next practices," those cutting edge initiatives that will lead to sustained competitive advantage in a changing world and marketplace.

Prahalad argued in this article that today's top management is too narrow in its focus to see "best practices" and must refocus their initiatives to uncover "next practices."

He advocates that top management ask ***six*** questions in this search for "next practices":

1. Is the problem widely recognized?
2. Does the problem affect other industries?
3. Are radical innovations needed to tackle the problem?
4. Can tackling the problem change the industry's economics?
5. Will addressing this issue give us a fresh source of competitive advantage?
6. Would tackling this problem create a big opportunity for us?

By way of example, we have considered previously how Progress Energy—the utility that serves portions of North Carolina and Florida—had launched a Smart Grid Program Management Office (SGPMO).

Why did they do this?

The logical answer was that they anticipated that the interest in Smart Grid would result in major project initiatives not only for their own organization but also for other contributing and supporting organizations such as General Electric. Smart Grid would touch the lives of not only the electricity community but also the customers and all stakeholders who have an interest in the application of SG technology. Smart Grid would be an evolving concept because "the more you learn about what you don't know about it, the more inclined you are to change direction with regard to subsequent actions."

In other words, the six questions that Prahalad proposed we ask of a problem facing an industry and its stakeholders would surely lead to competitive advantage for someone—why not Progress Energy?

So defining a Smart Grid PMO was a very "smart" thing to do when faced with all the unknowns about how Smart Grid would play out.

The same could be said of Apple creating a PMO for its financial organization, reporting to its CFO. The rapid product development that Apple has undertaken over the past twenty years or so has meant that its problem was how to mass produce sophisticated products that could still meet low cost requirements and a margin that would sustain Apple's viability.

Many organizations engaging in a "next practice" analysis will find that establishing a PMO will provide them with a long-term competitive advantage.

So what are we saying here?

Those of you who would seek to plan the next generation of PMOs need to look to the six questions that Prahalad succinctly stated in his HBR column. Finding that "NEXT PRACTICE" will become standard management practice if companies want to succeed.

IS IT TIME FOR A PMO RENEWAL?

Related to the idea of Next Practices is that of Renewal. I have always thought that summer was a time for renewal.

After all, isn't that why there was no school in the summer, and why your parents sent you to those summer camp enrichment programs? It was so you could relax and enjoy some subjects and activities that you were not exposed to during the remainder of the year. And, as you get older, perhaps you had a chance to attend a special summer program in a college or university setting, where other students like yourself could engage in debates and cultural activities that were enriching.

At some point in your life, you may have noticed that there was a transition away from those renewal activities that were planned by others to renewal activities that you planned by yourself. Maybe you didn't notice this transition since you were caught up in the fast pace of growing up, but, as John C. Maxwell has said, "Man has choices and the choices a man makes in turn make him."

So, Summer is a time for renewal. It's up to you to grasp those opportunities for renewal wherever they might be—in whatever season—in your life. Maybe, for you, it's the back-to-school aisles in the stores that get you energized instead.

If you are part of a larger project community or a member of a PMO, it might be time for you to seek out some activities to renew your PMO as well.

Give this some thought. How can I as a project or PMO practitioner provide some renewal activities for my PMO?

Show some leadership and initiative to make this renewal something that lasts year-round in your PMO.

For example, look at some liaison opportunities with a local PMI Chapter, or some special training that could be brought on-site to enrich your colleagues in the best practices of a PMO.

3

Project Leadership

HOW DO WE GET BACK TO "PROCESS" AS THE BASIS FOR OUR PROJECT MANAGEMENT PRACTICES?

When I was about ten years old, growing up in Concord, North Carolina, there was an activity that I really enjoyed each week. It was watching the "Baseball Game of the Week."

Now you have to understand that baseball on television then was very different from baseball on TV today. There was no WGN in Chicago to carry the Cubs games. There was no TBS in Atlanta to carry the Braves games. There was no ESPN to analyze the game of baseball from forty different angles.

There was simply one station—CBS—that carried whichever game it chose to cover, usually from the northeast and, most often, from Yankee Stadium featuring the New York Yankees. Go figure!!!!

Every Saturday, I was so excited for the 1:00 PM broadcast that I was actually seated in front of the television, with my favorite sandwich and beverage, at 12:30 PM waiting for the game.

As a result, I started watching a weekly series sponsored by the National Association of Manufacturers (NAM), entitled "Industry on Parade," that preceded the baseball game. The series was a showcase of the latest manufacturing technologies and new approaches to emerging industries, like plastics, after World War II.

Although they did not emphasize its importance in the weekly shows, "Industry on Parade" was a great introductory course in PROCESS. In fact, it was probably the best "course" I ever took in PROCESS, as defined by the traditional Michael Hammer: "an organized group of related activities that together create customer value." That includes my coursework while

DOI: 10.1201/9781003593645-3

earning my undergraduate and graduate degrees in science, engineering, and business.

What "Industry on Parade" did for me was to "imprint" on my mind a visual image of a specific set of activities that together led to a finished product, service, or goal. The experience had such a profound effect on me that my approach to almost any situation from that point forward was to look for the PROCESS behind the actual description of the event.

And so it was that, when I learned about the discipline of project management, my first reaction was to think of it as a PROCESS. Project management, by description, had a "baseline" or "as is" state. Any improvements to the way that the project management state was planned and delivered then became a PROCESS IMPROVEMENT to an established or "baseline" practice or discipline.

As I observe the way we try to add complexity and more competencies to today's project management practices, I wonder whether, in our haste to advance the practices of project management, we have forgotten the basic understanding that a project management "baseline" is first and foremost a PROCESS. In other words, it is a set of activities which, when carried out in a specific sequence, yields a planned result that can be forecasted, measured, and made repeatable and consistent. (See Hammer's book *Beyond Reengineering*, for example.)

Understanding a PROCESS is easy when the process is tangible, especially for visual and kinesthetic learners. Yet, practitioners of project management often seem to fail to see "the forest through the trees"—they fail to grasp that project management is, at its core, the implementation of a PROCESS. I truly believe that we should return to something similar to "Industry on Parade" for the training and guidance of our next generation of project managers.

Today, PMBOK characterizes itself as being a process or a series of processes which when taken together constitute "good project management practice." But it's not PROCESS in the purest sense.

Those who cite PMBOK as the process by which project management plays out are missing the key point that every PROCESS must have an objective, must have controls, and must be measured to ensure that it meets objectives and goals.

Those who develop the next iteration in the project management discipline should look back at "Industry on Parade" as a benchmark in creating long-lasting, instructive visual images that illustrate project management as a PROCESS.

A SELF-DEVELOPMENT FRAMEWORK FOR PROJECT MANAGERS

When I lived in Tulsa, I attended a weekly luncheon networking group, the Tulsa Business Alliance. During my participation there, I noticed that many people, including some members of my luncheon networking group, were reluctant to participate in the "social media" revolution, particularly for business purposes. This reluctance has continued through the present day. I believe the reason for this reluctance is that some people cannot visualize or relate to how they fit in or can participate in the "social media" phenomenon as we know it today.

As part of my networking group, I prepared a program titled "How Social Media Can Be Used to 'Obtain Information' or 'Provide Information' in Today's World." I wanted to identify and clarify some "roles" they could take on in the new social media environment. In essence, that is what social media allows us to do—to provide information to others and to obtain information for ourselves to facilitate our lives and our processes.

As a framework for the discussion, I chose an Engagement Pyramid or Model that Charlene Li presented in her book *Open Leadership: How Social Technology Can Transform the Way You Lead.* An Engagement Model describes the various "roles" a person can take on with regard to "subjects of interest" or "communities of practice." The social media this is directed at can be ".com" sites, Facebook, LinkedIn, Twitter, or others.

After the presentation, I reflected on the feedback and questions from the network group. It was then that I realized that this Engagement Model could also form the basis for a self-development framework for project managers. Was that a stretch or a valid observation?

Let's examine this Engagement Model by Charlene Li to see how it fits the bill as a self-development framework for project managers.

Certainly, project managers seek to obtain information and provide information to the larger environment at hand, and they are particularly interested in project management as a subject and the project community as a "community of practice."

The levels of Charlene Li's Engagement Model in ascending order are Watching, Sharing, Commenting, Contributing, and Curating.

Charlene Li's Engagement Model shows that, at the lowest level of engagement, individuals "watch" events and read materials on media sites as a means of understanding the environment and obtaining information on subjects they are interested in. Sites such as "projectmanagement.com" are particularly good for "watching" because they provide numerous articles on project processes, methodologies, project competencies, and so on. So, the first level of self-development for a project manager would be to "watch" and read many sources of project information in the media.

The second level of engagement according to the model is "sharing." In sharing project information through emails or other media, a project manager begins to decide what is relevant to the larger project community from his own experience, and he also receives information through sharing which outlines others' perspectives on project issues and competencies.

The third level of engagement is "commenting." It means that a project manager has read an article or a blog post or some reference material and wishes to comment on the content or material presented. It also means that the project manager has formed some "position" with regard to the particular subject matter and is willing to defend his position as feedback to the source providing the original media information. In this regard, a good example might be a project manager reacting to a PMBOK process as specified and, based on his experience in that area, wishing to provide some comment on the subject.

The fourth level of engagement is "contributing." Project managers who contribute to media information on projects or other subjects generally have broad experience in the pros and cons of actions in that area, the insights from experience in projects, and the willingness to put their opinions and positions on the line in online media for all to digest, ponder, and comment on. Everyone is an "expert" in some area of their discipline or interest, and contributing is a gesture of a project manager who wants to "go on record" with his observations and insights.

The fifth level of engagement is "curating." Charlene Li says that less than 1% of the population takes part in curating. Curating implies that the project manager wants to take on the role of setting standards for project management and processes, monitoring the content of social media in a subject matter area, and making judgments concerning the appropriateness of the material to the subject. At this level, a project manager must have a genuine interest in furthering the literature of project management.

As a self-development framework for project managers, the Engagement Model can be broadly interpreted in terms of "roles" a project manager can and should take on in his or her progression in the project community.

I hope this explanation has made it clear about how a project manager can pursue self-development using the Engagement Model. Certainly, formal coursework and individual study fit into any development plan for a project manager seeking to develop himself or herself fully in the project community. But the Engagement Model, if thought of in simple terms of how to provide information and obtain information in the media, can provide a very flexible and actionable plan of development for oneself in projects.

For project managers who are still grappling with this concept, I would suggest you use the ".com" site below to get started. "Projectmanagement.com" is the site for the online IT project community and provides an opportunity for each of the "roles" described above to be exercised and better understood.

For example, a project manager can join "projectmanagement.com" as a community member, read and comment on articles on the site, contribute original articles on project management subjects of interest, share information and insights through comments and blogs and answers to discussions, and generally create their own active roles with regard to the community.

As a self-development exercise, Charlene Li's model is an excellent way for project managers to progress from neophytes to skilled project practitioners.

WHAT WILL BE THE ROLE OF SOCIAL MEDIA (AND OPEN LEADERSHIP) IN THE MODERN PROGRAM MANAGEMENT OFFICE?

I remember a PMI Tulsa Chapter meeting in which we discussed the various generations that are presently in play in the workforce and the implications that this wide variation of working styles has for project work.

One project manager brought up his work with virtual teams on projects. While he had formed an opinion of his team members' expertise through their virtual collaborations via email and instant messaging, he found that once he encountered his team members on Facebook and understood their backgrounds and their interests, he formed a more informed opinion and a greater appreciation for their competencies. His team members' Facebook profiles allowed him to see them in the real context of their everyday lives, including their extended interests, passions, and purpose. The combination of the virtual team's collaboration and the team's interactions on Facebook allowed a more complete picture of the team's capabilities, which contributed to the team's current and future project success.

Several other roundtable participants, however, chimed in that they believed that the interest in "social media" would die down and be less of an influence on project work. There were about as many different opinions on the role of social media in project management and the modern Program Management Office (PMO) as there were people at the table.

I am certain all of us have had similar discussions because the project community—and indeed, all of society—cannot escape the fact that openness now rules the collaborative nature of our daily interactions and our lives, especially as virtual work environments become more common.

The role of social media and "openness" continues to unfold every day with new forms and branches of social media being utilized by the general public in a variety of ways. Every year, new and additional forms of social media and other team-building or messaging applications find their way into more facets of our everyday lives. Embracing them is a necessity, just as learning to use smartphones was fifteen years ago, and learning how to use Zoom was during the Covid-19 pandemic.

So, what will be the role of social media (and open leadership) in the modern PMO? Well, it depends.

It depends on the mindset of your PMO's leadership with regard to the use of **any** form of media to connect your PMO's employees and processes with the external environment and the PMO's stakeholders.

Roger Martin, Dean of the Rotman School of Management at the University of Toronto, has stated the problem in this way:

> The struggle in balancing openness and control is a universal human problem. While most leaders agree that greater transparency and authenticity can lead to significant benefits, many remain paralyzed by the risks involved in opening up the lines of communication with their stakeholders. Tapping into the power of social technologies isn't about mastering the latest shiny technologies, but instead having a clear idea of the relationships you want to form with your stakeholders.
>
> (From the introductory pages of ***Open Leadership*** by Charlene Li)

How the PMO's leadership sees itself in this new social environment is very important to the types of decisions that the PMO makes as to how it will employ these new social technologies. In fact, in ***Open Leadership***, Charlene Li of the Altimeter Group has used the phrase "Social Leadership" to focus on how organizations will develop to handle these new social technologies.

Social leadership is defined as having the confidence and humility to give up the need to be in control while inspiring commitments from people to accomplish goals.

First, if your PMO is still in the mode of looking at itself as a cost center—as many IT PMOs generally see themselves these days—then your approach to social media will probably be to see it as a drag on employee productivity and a cost increase item due to the linkage with the external environment and stakeholders.

But, if your PMO sees itself as a strategic partner with the other business or functional groups in the organization, then you will probably approach social media as an "opportunistic" and value addition vehicle.

The fact is that the PMO is like any organization. It develops a culture over time which, in essence, is "the way things get done here." Key processes constituting organizational culture are information and communication flow, decision-making and authority flow, and human resources flow.

Organizational cultures tend to exist on a continuous spectrum between "closed" and "open" systems. Once a culture is established, it is difficult to change.

This observation comes from experience.

When I worked for ConocoPhillips, the corporation was a member company of a group known as the Information Technology Research Institute (ITRI) of the Walton School of Business at the University of Arkansas. Other typical member companies were Tyson Foods, Dillard's, WalMart, Federal Express, J.B. Hunt, and Dell.

The ITRI "member companies" sent representatives to meetings at U of A several times a year to discuss Information Technology and Project Management issues as they impacted the work of the various member companies. I was often a participant in these discussions and even facilitated a PMO Roundtable meeting.

Each year, the ITRI identifies a "Top Ten List of IT Issues" which they believe will impact the member companies' IT and business strategies in the near future. It is always interesting to look back at how this list has changed over the last ten years to see what hot topics have risen to the top of the list. Project Management and PMO operations, for example, have steadily risen on the list. Innovation and Business Intelligence have appeared and moved up the list.

Consider some issues discussed in this forum:

> The explosive adoption of tools such as Twitter and Facebook have created new modes of personal and mass communication in the last 18 months, but business remains unsure of its benefit within the corporate walls. Citing potential productivity decreases and the need to control corporate messaging, many companies have adopted a wait and see approach to this rapidly evolving phenomenon.
>
> (From the ITRI 2008–2009 Annual Report)

Now, keep in mind that this is a group made up of member companies who have **embraced** the evolution of the use of the PMO as a key delivery tool for project value to the organization (not all organizations will have done accordingly at that point in time). Today, of course, there is no question that serious portions of corporate messaging and marketing, not just by start-ups but by the largest of established businesses, take place on social media platforms.

Professor Puneet Manchanda of the Ross School of Business at the University of Michigan conducts ongoing research in social interactions, social networking, and social media.

Manchanda recommends that companies pursue both "active" and "passive" strategies with regard to social media. In the "passive" mode, organizations will gain valuable information about their particular subject by monitoring what others have to say about their day-to-day tools, methods, and practices.

For example, the evaluation of new tools, methods, and practices for Project and Portfolio Management will be played out in social media by those whose interests are in either (1) using the tools, methods, and practices in their own business context or (2) those who wish to advance the use of such practices in the community at large. A PMO can gain valuable insights by merely monitoring the conversations taking place in that community and taking action accordingly.

Alternatively, an organization can take an "active" role in social media as it shapes the landscape for tools, methods, and practices of its discipline by taking particular positions that would advance its own particular way of doing business in its culture and context. The various "stakeholders" which a PMO serves could then be kept in tune with the latest thoughts and actions of the PMO and, in many cases, could be asked to participate in the development of new tools, methods, and practices.

This would be a powerful approach; it would build that all-important element of commitment which every PMO needs from all stakeholders.

Charlene Li, in *Open Leadership*, identifies **four major themes** in the transformation of an organization such as a PMO to an open, social media organization:

1. **Values drive the vision**
2. **Leaders set the tone and the example for others to follow**
3. **Extending the old culture into the new**—if culture is made up of norms and values, the organization must define new processes to define how these relationships will work.
4. **Systems and structure sustain the transformation**—supporting the new culture are new incentives and recognition systems, as well as revamped processes and procedures that govern interactions both internal and external to the organization.

Let's take **two examples** of how this might operate in a PMO which "actively" uses social media.

First, as is set forth elsewhere herein, I advocate gathering Project Lessons Learned at major break points in the project or, as some of you might say, at various stage gates. The project manager could solicit project feedback from team members, sponsors, and other stakeholders using social media such as Twitter. A key to this change in the business process would be how this feedback is collected, reviewed for common themes, documented, and shared with all project participants.

Second, let's assume an example where a design point has been reached in a project where a potential scope change might occur depending on the tradeoffs and decisions that need to be made jointly between the project team and the stakeholders. Once again, social media such as Twitter could be used to "inform," "solicit feedback," and facilitate the decision and sign off on the resulting scope change if necessary.

You can see how both these examples require certain norms, values, and beliefs to be supported by and in turn support the essential business processes.

Trust and commitment are assumed ingredients supporting the scenario of Open Leadership.

Are there organizations that we can examine as case studies on how to structure processes, incentives, rewards, values, and beliefs in order to support the new open reality of collaboration?

Cisco and Proctor and Gamble are two such organizations cited in Charlene Li's book that project managers can study for clues to our own PMO transformation.

Make no mistake!!! This does not have to be a major upheaval of existing processes and relationships. But its success does rely on trust, authenticity, and a willingness to approach control of key organizational processes in new ways.

This will certainly be a topic for discussion for many years to come as PMOs and social media tools and approaches advance.

TECHNOLOGY IS THE KEY TO YOUR SUCCESS AS A PROJECT LEADER

I was watching a Georgia Tech men's basketball game when the ball was deflected out of bounds, and the official signaled that it was in the possession of the visiting team. He was quick to go to the scorers' table and check the video replay. The replay showed clearly that the official's original call was in error. What did this demonstrate?

1. Technology was the key to support the rules of the game.
2. The official was comfortable in relying on technology.

So, you are asking as a project management professional, what does this have to do with me?

Everything. As a project leader, you must leverage your use of technology to be successful in project situations.

But this is nothing new. Absorbing and effectively using technology is essential to staying current and competitive in project management.

Project collaboration software and data science applications are two of the most dynamic and useful technologies in project management. The application of Benchmarking and Best Practices has become widespread to set teams apart from the competition.

As a project management professional and a leader in your industry, you can leap to the head of the class in the innovative application of new technologies. Measure your success at this by how many requests you get for sharing and referrals.

A lot has been written and said over time about first-mover advantage in market moves or new technology implementation. As a project professional, you must always be prepared to lead in the face of change. Good luck!

WHAT CAN I DO? REMEMBER THAT YOU ARE AN "EXPERT" IN SOME SUBJECT

Our need to participate in the larger community, whether via social media or otherwise, is felt strongly in the wake of tragedies in our communities. After a tragedy, many people ask, "What can I do to help?" Those who are too far away to provide direct help feel the need to get involved either in the specific situation or the larger debate, on what can be done to avert future tragedies of this type. Too often, we feel like our skill set isn't an obvious fit for the types of activities that often follow a tragedy, like caretaking, fundraising, or lobbying, for example.

Step back for a moment and realize that you too are an "EXPERT" in some specific scenarios you have lived through or have witnessed others reacting to in their own "communities of practice." Everyone is an EXPERT in some subject.

When you ask yourself the question "What can I do to enrich the discussion or questions about life experiences?" know that you are an EXPERT and can provide meaningful insights that others will value. The trust they have in your advice will be built over time as you continue to step forward to offer more answers to questions. Your insights will gain acceptance by a broader group of acquaintances as you continue to offer your EXPERT opinion.

For example, management and organizational skills are often exactly what an organization or group needs to leverage its efforts to achieve real change. Your skills can be scaled up (or down) to fit the needs of the community. Don't be afraid to participate in your community and lend your expert (yes, expert!) skills to the task at hand.

Don't hesitate to start this process today. The quicker you get positive feedback from taking this action, the more you will feel involved in helping others cope with human experiences. It's your opportunity to give back and to say "Here is what I can help with."

By way of example, here's a personal story. I launched my consulting business in January 2010. At that time, I wanted to focus on teaching and sharing with the "project community" some of my insights gained from at least fifteen years working in PMO and Project Office type environments. The experiences I gained working with very capable and innovative project managers and PMO staff, I believed, would provide some great direction for those in the project community who were seeking answers to personal questions about project management. I realized that everyone at some time is seeking advice from knowledgeable sources of information in specific subjects or disciplines. That is just a fact of life.

Encouraged by my daughter, to accompany my consulting business, I decided to write a blog on PMO structure, activity, behavior, development, and performance. I needed a "name" or a "byline" to give it some authenticity.

My daughter suggested that I call it "MEL BOST PMO EXPERT." Although at first I was a little concerned about using the word "EXPERT," after discussing with my peers in the project community, I realized I was truly an "EXPERT" in certain specific PMO situations and scenarios. That is not to say that I knew everything about a specific subject or how to react to specific actions and behavior of project managers, it was intended to provide those in the project community with some trusting advice about their very own lives as project managers. It is true that our own internal self-talk can be the most defeating! This is why we always have to surround ourselves with a community that will support and encourage us.

Have confidence in your expertise and use your expertise to participate in the project community and to help others. These are our unique skills to share with the world, and nothing is more fulfilling.

WHAT IS THE FOUNDATION FOR YOUR PROJECT TEAM LEADERSHIP?

In 1985, Atlantic Richfield Company (ARCO) moved me from Chicago where I had worked in its R&D facility, to its headquarters in the ARCO Plaza in Los Angeles, to head up a strategic planning group for ARCO Products. The day I arrived to find my new office was really exciting. As I walked through the corridors of my new floor, I happened to see a paperback book on the desk of an empty office. In the haste to vacate an office and discard old materials, someone had left this book for a new occupant to find. I picked it up and skimmed it quickly. It was *Quality Is Free* by Philip Crosby, which he wrote in 1979. Being the book hoarder that I am, I took the book with me for future reference. I had no idea what value it would offer, but I knew that impulsive finds had always benefited my learning and ability to evaluate new opportunities in some way.

Over the years, as I worked in project management and process improvement, I learned that the Crosby book had been the foundation for Watts Humphrey at IBM in 1986 developing what came to be called the "Capability Maturity Model." Crosby's "Quality Management Maturity Grid" had been used as a model for software engineering and development and the ways in which organizations move from "ad hoc" processes to fully optimized operations.

The book that I had picked up casually in 1985 was a foundational book in the sense that its concepts formed the basis for many significant models and frameworks in all phases of business, engineering, and technology.

As a project manager and team leader, what are the foundational concepts upon which you build your daily processes and standards? Have you ever really reflected on this question? Too often it is easy to fall into habitual activities that seem to fit the scenario you are facing at a given time. Habit is the fallback position when casual day-to-day decisions occupy most of your time.

Being project managers means we traditionally follow the practices of project management that are laid out in PMBOK and include resource identification, scheduling, task definition, milestones, and execution, among other project concepts. That is a given.

Foundational concepts often arise from values and beliefs. How to handle expectations, dilemmas, conflicts, negotiations, allocation, priorities, stakeholder expectations, sponsor rights, ethical issues, and project team makeup are among the usual day-to-day tasks that occupy your time and attention. As a project manager, you are a role model for your project team. Your behavior often foretells how your team members will behave in the future when faced with similar situations.

How your team handles setbacks, unrealistic goals, performance shortfalls, and other human events can also be affected by foundational principles that you, as the project manager, can anticipate and manage.

So, project management is much more than guiding the planning and execution and closing of a project. It provides guidance based on values and beliefs that you put in place and which will become the standards by which your organization is judged as a PMO. Are you prepared to evaluate your performance on these foundational principles and to make adjustments if necessary?

If not, find another line of work. Real project management demands no less than your best effort to create the foundational basis for success.

By the way, I still have *Quality Is Free* and often refer to it as a foundational text in my personal library.

THE TREMENDOUS IMPACT OF ROLE MODELS ON PROJECT MANAGEMENT LEADERSHIP

When I was a teenager, I walked into our family room one evening and found my father on his hands and knees in the middle of the floor. Around him were four flip charts, and he was drawing with a crayon freehand on each of the four flip charts.

"What are you doing?" I exclaimed.

"You remember last week when I was in Detroit for four days on that Service Training trip?"

"Yes. I wasn't sure which topics you were covering, but does this have any relation to that?"

"Yes. I am preparing some charts to explain the new transmissions we are introducing in our tractors to our service managers and our field organization. This is really exciting stuff, but it takes a picture to make it come alive."

"Didn't the company give you any material to train your service people with; pictures and descriptions of the new transmission?"

"Well, of course they did. But I did not think that those handouts really conveyed the ease of operation and the full leverage that exists with the new transmission. That is what impressed me the most."

As he proceeded to fill the four flip charts with diagrams and words, I began to see what a "trainer" my father really was. I never associated his job as a Service Manager in the southeastern District of Ford Tractor with a training job, but he clearly did.

I was amazed that, by the end of the drawing that evening, when he explained the new transmission and its features to me, I actually understood what he was saying. His pictures and words combined the right thoughts and pictures in my mind to make sense of this new engineering feature.

Now, I have to reflect on this situation today. Is this why I am a good trainer? Is this why I use words and pictures and diagrams and flow charts to express my ideas and concepts to my audience? Did I see someone else in action who I copied in my later life as I developed into an excellent trainer?

Yes. Decidedly, I believe that role models are the most effective means of experiential learning that we can have. The lasting outcome of having role models in our lives is to provide living, responsive models for our development and our maturity.

I had never really considered my father to be a trainer before. He was a manager of a number of other individuals who reported to him in the organization.

There are several principles at work here. Lou Tice provided these principles in his teachings on developing potential and reaching for your own heights of success.

First, people move toward and become like that which they think about. Having role models to view and interact with brings us closer to action that resembles the role model's actions and outcomes. That is not a literal statement, of course. It really means that we imitate much more than we would think we might in our lives.

Second, people act in accordance with the truth as they perceive it to be. When a role model presents a "truth" to us that matches our own value systems and beliefs, we adopt the mannerisms and processes that the role model exhibits.

What do good role models do that set them apart from ordinary people?

I believe that role models set the standard for the way a discipline or a field of endeavor is carried out.

- **First**, role models are passionate about the field of their discipline.
- **Second**, role models do more than the minimum required to get the job done. They have a passion for conveying to other people more than exist on the surface. Role models are often "risk takers," but "*calculated* risk takers." They often rise to leadership positions in groups.
- **Third**, role models are not pretentious. They don't try to show how much more they know or how much more they can convey. They are "genuine" in their approach and that "genuineness" shows through to others clearly.
- **Fourth**, role models view "teaching" as part of their job, part of their mission in their field.
- **Fifth**, role models can provide extremely effective experiential learning for project managers. Because role models can provide feedback and answer questions about their actions and resulting outcomes, they can provide the perfect "simulation" for viewing.

With the exception of mentoring—a one-to-one relationship between two persons—more than any other type of learning—classroom, team experience, case studies—role models are the most desirable of all leadership training routes.

Let me provide a few examples of good project manager role models from my experience.

There was a project manager with whom I worked at ConocoPhillips whose assignment was to merge three companies occupying the same geographic region into one company that used SAP as the financial and management reporting system for the company. Before he assembled a project team or talked to the other principal stakeholders in the merger, he traveled around to all the company locations and talked to each employee individually getting feedback and comments on what had worked well in the past, what processes should be maintained, what processes should be abandoned, and how each person felt about their role in the new company to be formed. At the time, some IT Shared Services Management questioned whether he should not have taken the trip, but rather, whether he should have got on with the task at hand of organizing his project team since a deadline loomed in the not-too-distant future for the merged company to be active. What he did, however, was to successfully gain commitment from everyone at all levels of the organization, while reassuring them of their place in the new organization. A real WIIFM (What's In It For Me) example, so to speak. When he formed his project team, he included many people from each of the three companies who advised and commented quickly on the actions of the project team. Everyone saw this as a great example of a role model at work solidifying the success of his project and the company going forward.

Another example is my introduction of Project Lessons Learned Breakfast Forums to the IT Shared Services PMO organization. We had not had a formal Project Lessons Learned process in place to that point in time. I knew that commitment to such a program would only be solidified if I gained the confidence of a few "risk-taking" project managers to make it happen. One project manager stepped forward and volunteered to be the first project manager to address the Breakfast Forum with her Project Lessons Learned. We videotaped the sessions, and they became a hit on our intranet. The project managers who stepped forward and participated in the program became instant role models for other project personnel.

What has your experience been with role models? Perhaps someone has influenced your actions and behaviors without you really realizing the effect they have had on your thinking and your actions.

Look for role models in every discipline you are involved in. They are worth the time and effort. They are the essence of the new project leadership.

IN THE EVOLUTIONARY STAGES OF A PMO, HOW CAN YOU IDENTIFY THE LEADERS WHO WILL ENSURE THE SUCCESS OF THE PMO AT STEADY STATE?

There seems to be an abundance of recent research and published information on leadership in organizations these days. Ironically, however, during the same time that the old method of listing and evaluating "attributes" of leadership has yielded to the critical examination of actual behavior in significant leadership scenarios, there have been fundamental breakdowns in basic leadership in some highly respected public and private organizations. Much of the recent research and writing has focused on these breakdowns and their underlying causes.

Many of us have observed both good and bad leadership in evolving PMO settings, and I am sure we have stopped to think "Now, what could have caused that particular outcome … especially when it seemed that everything was moving according to plan for that project or program?" In order to avoid such breakdowns in your PMO, let's consider which criterion contributes to the emergence of key leaders in the PMO. In particular, I want to address how, in the evolutionary stages of your PMO, you can identify the leaders who will ensure the success of the PMO at a steady state.

Now, please understand that achieving a "steady state," in the sense of systems and process, may be an ideal situation that each of us hope to achieve for our PMO. In actual fact, change will always be the norm, and leaders will have to respond to "change" in innovative and insightful ways.

In my experience working in an IT Project Office, and in a PMO and helping develop several other PMOs from "grass roots," I have had the opportunity to work with some remarkable project managers and PMO infrastructure managers at the inception stage, the developing stage, at partial maturity, and at full-blown maturity. As a result, I have had a chance to compare my observations with the emerging leadership literature. Make no mistake; these observations are not intended to be exhaustive and complete. But they should help us to develop a PMO Leadership Model which we can continue to build on as we collectively view more behaviors in PMO settings.

Here is a list of characteristics of emerging leaders in a PMO:

1. **Calculated Risk-Taking Attitude**: In my experience working with PMOs, those project managers and project team members who displayed a risk-taking attitude in two key areas were most often the individuals who rose to leadership positions. First, those project managers who volunteered for the high-risk projects because of a "can do" attitude were often the emerging leaders. Second, when I was directing the development of Project Lessons Learned for a Breakfast Forum in which the project manager addressed the PMO project community about some key lessons from his or her project, those project managers who stepped forward and volunteered for the Breakfast Forums were most likely the ones to rise to leadership positions. In fact, the first five project manager volunteers became leaders and formal group managers. So, if you are a PMO developer, look for the "risk takers," while recognizing that I am talking exclusively about calculated risk-taking.
2. **"Commitment" vs. "Compliance" Perspective**: In every PMO, and especially in those that are in the startup mode, there is a range of attitudes toward the mission, vision, and values of the PMO organization. At one end of the spectrum is "compliance"—the condition in which a person complies with the processes, standards, and procedures of the PMO but doesn't really buy into the overall mission. These individuals most often come from "operationally" oriented positions which relied on "making a daily list of things to do." They often developed a work plan for a certain period of time which was rather inflexible in its components. On the opposite extreme of the spectrum is "commitment"—a condition in which a person is totally committed to the mission, vision, and values of the PMO even though he or she might not be the most competent project manager in the organization, and even if the direction of the PMO has not yet been established with much certainty. In my experience working with an IT Project Office and several PMOs, those persons who were "committed" and who demonstrated that "commitment" were most often the individuals who rose to leadership positions.
3. **Effectiveness through Dialogue and Influence**: In every organization—at all levels of the organization—there are individuals who distinguish themselves by being extremely effective at getting things done through Dialogue and Influence. These are exactly the people who were the object of research by VitalSmarts (which resulted in books such as

Crucial Conversations, *Crucial Confrontations*, and *Influencer*). They are the same people that William Bridges talked about in his book *Job Shift* who, when faced with rapid or unrelenting change in their organizations, developed the ability to successfully take on the roles requiring negotiations, brokering, translating, collaborating, facilitating, and so forth.

4. **Performance Focus**: In every organization, there are individuals at all levels whose attention to the management of their own performance—and that of their teams—goes beyond what is required of their individual organization's formal Performance Management Process. These individuals typically set realistic goals and objectives and measure the outcomes at intervals, making course corrections where necessary based on feedback and metrics. These people are conspicuous and stick out in every organization. They are authentic and genuine about their beliefs in performance management. And they rise to leadership roles in a PMO setting.
5. **Formal vs. Informal**: We have all recognized for years that the most visible structures in organizations are those organization's "formal" structures. Organizations are made up of organization charts, processes, standards, and procedures, as well as the underlying principles of efficiency, scalability, predictability, controlling influences; clear, disciplined, hierarchies; and rationality. However, some recent research by Jon Katzenbach and Zia Khan, entitled "Leading Outside the Lines," has identified an "informal" character of organizations that may be just as powerful at influencing organizational performance as the "formal" aspects.

The "informal" consists of loosely defined networks and communities of individuals with like interests. These communication and information flow mechanisms are far from formal; rather, they are adaptable, local, innovative, ambiguous, spontaneous, collaborative, and emotional. Individuals who know how to tap into the "informal" structures in the organizations can be leaders too. Elsewhere herein, when discussing the importance of communication, I have mentioned a very successful project manager of a very large SAP project who recognized that his project was integrated tightly with two or three other projects that were scheduled to be completed just before his project. He spent a considerable amount of time ensuring that those other projects had the resources and guidance necessary for their completion because he recognized the invaluable nature of their input and information flow to his own project. Now, he could have relied on formal Project Review Meetings to stress to all concerned the importance of their delivery, but instead he focused on the "informal networks" of cooperation and collaboration to gain "commitment" from everyone that all projects would be successfully delivered. That effort showed true leadership skills.

You will immediately notice that some of these characteristics pertain to project competency, and some pertain to business and personal competencies. This is consistent with Project Management Competency models such as the Boston University PM Competency Model, which stresses not only PMBOK competencies but also personal and business leadership attributes.

As you look around your PMO, notice the behaviors of project managers and team members who are "authentic" and "genuine" and who live the values of the PMO every day. Through good project experiences and difficult project experiences, there will be many who stand out in the crowd and become the ongoing leaders of the PMO.

PROJECT MANAGERS, ARE YOU READY TO ANSWER THE "CALL TO DUTY" IN YOUR ORGANIZATIONS?

Imagine a project manager who had been managing a major project for a large PMO for about nine months. The project had a two-year duration and was slated to provide new functionality and systems for a major revenue component of the company.

One morning he arrived at the office to hear the news that a Program Manager who had been managing a major effort that linked three projects into a program was leaving the company to join a major competitor.

Jack had often thought "What would I do if offered an opportunity to take over a major program in this company? I am not sure I am ready, but there may be little time to think about it if the offer comes."

Jack called his wife to tell her of the news and to discuss the ideas he had considered in making such a move. Of course, someone else might get the opportunity so he decided not to get too excited about it. They had often talked about such a "break" and what he might do in case it came, but they had never really done much planning because things always seemed so stable in his company.

Jack's wife told him about a story she was familiar with regarding Leonard Bernstein, the great conductor of the New York Philharmonic Orchestra.

She emailed him a link to the story from "America's Story," a collection of stories about the lives of great Americans. The story described how young Leonard Bernstein got his "big break" when he, at the last minute, was asked to conduct the New York Philharmonic's performance when the lead conductor was unexpectedly unable to perform. Bernstein's performance was so exceptional that it made the national news. The story posited that perhaps Bernstein performed so well in that extremely stressful situation because his passion for music overcame his situational stress.

After reading the article, Jack thought to himself: "Do I have the passion for project management that Bernstein had for music?" He reflected back on some of the crises he had handled in his project assignments and his desire to excel in this field.

Suddenly the office phone rang. Jack picked it up. On the other end was the Vice President, Program Management and Strategy Office for the company, who asked Jack to speak with him at two in the afternoon to talk about a new assignment.

Jack, like Leonard Bernstein, let his passion for his field give him the courage he needed to answer the call and step up in his organization. When you get your call, do the same.

FIVE TANGIBLE THINGS ASPIRING YOUNG PROJECT MANAGERS CAN DO TO ENHANCE THEIR CAPABILITIES

I have noticed recently, in the Discussion sections of LinkedIn's Project Managers, PMO, and PMO Bloggers Groups, that aspiring young project managers have been requesting advice from more experienced project managers and PMO leaders as to what steps they can take to enhance their ability to move forward in the "project community" and to advance their project management skills.

From my point of view, having worked in several PMOs and IT Project Offices, as well as having assisted in the setup of several PMOs, I would like to offer the following advice, which follows from something I read by Jillian Michaels, the very successful coach of the hit TV show "The Biggest Loser."

Jillian once said that individuals and organizations can accomplish anything they want to accomplish as long as they are **sufficiently motivated and have the capability to succeed.**

My advice is focused on the aspect of "capability" because there is much that aspiring project managers can do to enhance their capability. **Capability**

means educating oneself and understanding the business context within which they are operating.

By aspiring young project managers, I actually mean project team members who are qualified to move up to managing projects but who haven't found that right project to manage yet OR project managers who have just assumed a new project management assignment for which they must prepare themselves to fully succeed.

Here are five areas in which aspiring young project managers can enhance their "capabilities." These five areas will leverage your understanding of the project process, as well as increase your stature among other project managers in the "project community."

First, volunteer to record the minutes at an important project status meeting. Be sure to find or create a template that captures the participants, the date and time of the meeting, the agenda, the actions agreed to, the assignments for each participant as well as any completion dates, any issues that were not resolved, and the date and time of the next meeting. Too many project teams lapse into keeping very poor documentation of their meetings and, therefore, accountability for completing assignments is lost. By volunteering for this assignment, you will provide a key document to the project team from which they can build throughout the project.

Second, at the conclusion and documentation of the project requirements, survey the sponsor and the stakeholders to understand that they really know what requirements they have signed up for and the "commitment" that is required to carry out the project. By doing so, you will have input into any scope change that may arise ahead of the actual requirements modification. This will increase your understanding of the project process and your stature among other project managers.

Third, at the conclusion of the first major phase of the project, insist that the project team hold a "lessons learned" session. They will thank you later even though you may have to drag them to the table kicking and screaming to participate in the first session. So many things get uncovered when you "shine a little light on something" or focus on the actions and results at this point. Famed UCLA basketball coach John Wooden once said "It is what you learn after you know it all that counts." This has terrific application in Project Lessons Learned.

Fourth, find a key issue that seems to divide the project team members or the stakeholders and follow that issue by recording key actions and decisions made by the team and the stakeholders. At an appropriate point in the project, when it appears that the project team or the stakeholders or both have reached an impasse or a stumbling point, pull out your summary of the actions and decisions and review it with the combined project group. Some may not appreciate being confronted with the facts, and others may disagree with your facts or your interpretation, but no one can disagree with the harmony that will result down the road when they begin to see each other's different perspectives and viewpoints and the fact that a rational, realistic observer brought these viewpoints to their attention. You may have won yourself a job!!!!

Your facilitation of a bumpy issue by offering extraordinary insight and analysis is the key. Analysis is the essence.

Fifth, practice the four communication mechanisms that I discussed in one of my previous blogs:

1. Situational
2. Metaphorical
3. Empathic
4. Resonance

I use the word "practice" here because it will take the aspiring new project manager some time to understand the situations in which these four communication mechanisms will be most effective. But, they all have a place at some point in a project. By using these communication tools, your effectiveness as a project communicator will be enhanced manyfold.

By the way, you won't find this advice anywhere else—it is unique to taking a holistic view as to what a project is all about and to understand how stakeholders impact projects.

Good luck to you aspiring project leaders. Yes, you are and will be leaders if you follow this advice.

IS YOUR PROJECT IN 2-D OR 3-D?

Several years ago, on "The Tonight Show," Paul Reiser was a guest.

You may remember Paul as the male lead in the sitcom "Mad About You" which starred Helen Hunt as his wife. The madcap comedy of this twosome kept me in stitches for many evenings.

Paul was appearing on "The Tonight Show" to promote his new sitcom "The Paul Reiser Show." However, as luck often has it in the TV business, Paul's show was canceled after just two shows on air.

The way he said the TV executives pitched the cancelation to him was "Well, you have some viewers and then you have not so many viewers. Your show fell in the latter category."

So he was a "lame duck" in a sense but, since he was already scheduled for Jay Leno's show, he showed up to provide some laughs and a look a reality in the wacky world of TV sitcoms.

Jay asked Paul about his two sons, who I believe are seven and ten years of age. Apparently, the last time Paul and Jay spoke, Jay recalled that the two boys had requested a 3-D TV for their home.

Paul explained that one of his sons had exclaimed "Dad, didn't you see that last jungle picture in 3-D? The trees were slapping me in the face as the animals swung through the trees. It was so real."

Paul explained that he had always been a 2-D person himself and, in fact, throughout his life, he had tried to avoid being slapped in the face by anything resembling tree limbs as he navigated the world.

But, of course, the boys were different. They really enjoyed the idea that they could actually experience being slapped by something unreal and yet so lifelike in appearance.

How do you treat your projects? Do you go through the motions, treating your projects like 2-D, always hoping that something won't leap out and slap you in an unexpected manner?

Or do you approach your projects with a 3-D mentality, relishing the idea that every slap in the face represents a new opportunity to lead?

If you are experienced enough to have managed several projects, you know that 3-D is the only flavor for projects in today's fast-paced project environment.

So why avoid being slapped in the face by treating projects as 2-D? Take a risk and embrace new experiences and new opportunities to lead—you will be glad that you did.

THE ROLE OF PROJECTS IN SMALL BUSINESS

When small businesses are formed, the owners are most often concerned with day-to-day operations. If a small business was launched to provide a product or service, the daily maintenance of that product or service is the chief interest of the business team. The financial viability of the small business is dependent on the immediate revenues generated by the day-to-day operations.

Having a clear vision of what the business wants to accomplish is so important to the success of the business for a long run. A key ingredient of the Strategic Plan is an assessment of the Strengths (S), Weaknesses (W), Opportunities (O), and Threats (T). Vision and Mission are the results of the SWOT Analysis.

Day-to-day operations are certainly important in defining the business processes that will deliver the products and services immediately. But small business owners always ask the question "Where does my Strategic Plan lead me?" The answer is that the SWOT Analysis from the Strategic Plan leads to how the business sees itself in the future. In order to create that future, "projects" are identified and employed to turn "strategy" into "action." Projects are those longer-term efforts that develop plans, schedules, resources, and milestones that will help the organization achieve its future state. Inevitably, projects are longer-term in nature and require leadership and vision to turn the SWOT into reality.

You don't have to be Certified in Project Management to use project management principles. The project management framework is the application of structure and discipline to know when things are proceeding as planned and when they need to be recalculated, replanned, or rescheduled. In fact, a simple project management application was recently introduced known as "Monday.com." As your business grows, if you start to do business with municipal or other governmental entities, or projects involving permitting large dollar projects, or third-party vendors, however, your business counterparties may require you to have a Certified Project Manager (PMP).

A NEW REVELATION: NOW YOU TOO CAN TAKE ON A PROJECT MANAGER "ROLE"

One of the fallacies that many people who do not work in the project management field believe is that all those people in the project community are avid students of project management and that many of them are "certified" by somebody to be "real" project managers. In reality, it is a fact that anyone can use project management techniques and disciplines to improve their work processes or their leisure activities as well.

Take for example the TV channel HGTV (Home and Garden Television). This channel is focused on home ownership and all that goes along with it: acquisition of homes, marketing of homes, improvement of homes, and so on. It is no surprise because home ownership is so important to economies, and a home is probably the largest single investment of a person or a family in a lifetime. So it is natural that HGTV focuses on all types of "situations" that involve homes. I have noticed that over the past five years, many of the programs have begun to incorporate project management techniques and methods in their stories and in the narrative.

You too can be an active project management role player in your own home situations.

As you watch episodes on HGTV, look for these distinct topics:

1. Requirements Management—The focus here is on what is the intended outcome of the project. It might be a new patio, garden wall, or a restored room in a home. Watch HGTV as the project teams try to pin down exactly what they want to accomplish and then see if the clients or customers introduce any "scope creep" as the project proceeds.
2. Scheduling—This is a necessary piece of the home improvement since it determines how quickly some actions can take place and what actions must be completed before any new introductions of features occur. For example, suppose an existing floor needs removal before a new tile floor is installed. There is a sequence to this that helps determine the schedule.
3. Resources—How many people are going to participate in the project? What physical resources such as paint and draperies will be used? Identification of these resources is essential to a good project management plan.
4. Scope—It is as important to determine what will be included in the project work as it is to determine what will be excluded. Documenting these in some manner is a good idea to keep reminding the project team what they are working on and how it relates to the resources allocated to meet the objectives of the project.
5. Risk Management—Although perhaps not a prime consideration for new project managers, it is important to keep in mind what can go wrong and what events might dictate a change in plans or introduction of an alternate plan if certain events are triggered in the course of the project.

As you watch episodes of HGTV, look for these elements to be introduced to ensure that the project is completed on time and within budget with the right requirements being satisfied for the project. Try out some of these topics in your next initiative to see how smoothly you can complete the initiative.

Yes, you have stepped squarely into the project management arena and, if you continue to develop your expertise with regard to these topics and others that arise in HGTV episodes, you will continue to develop your capability to approach any new project with real project management tools and practices.

THE REWARDS IN SHARING YOUR SECRETS OF YOUR STARTUP BUSINESS

If you are a small business owner or an entrepreneur ready to start a new venture, you've heard all the opinions about whether to share your secrets and your uniqueness with the business community you are part of. On the one hand, the opinion seems to be that you should not give away what your customers and stakeholders should be buying from you. On the other hand, sharing with your customers and stakeholders creates a connection from which they can really understand your expertise and what you have to offer them.

Based on my experience, I reside in the camp that advocates sharing with your business community. There are very few business spaces that are so narrow that there is only room for one business in the space.

WHERE ARE THE RESOURCES TO FINISH THIS PROJECT? CREATIVE PROJECT LEADERSHIP

At some time, every project manager, from those in the sandbox to those in multi-million-dollar projects, has asked the question: "Where are the resources to finish this project?" It is inevitable in our lives that something causes our projects to run short just when the end is in sight. Call it poor planning or poor management, things happen.

When this happens in large organizations, we typically look around for another project that is doing well to borrow from to alleviate our dilemma.

How this is handled is usually a matter of how the organization divides its resources and allocates its funds.

But is that the first course of action real project managers should take?

My answer is no. It may be "pie in the sky" to look at the situation this way, but every project manager should look internally first to find the solution.

And of course, you are going to say "Get real. Money doesn't grow on trees."

Louis Tice, leadership guru and founder of The Pacific Institute taught his followers that the energy to find the resources that aren't there is internal to each person who seeks it. He taught us to develop bold plans for what we wanted to create and then turn them over to our subconscious and our minds to act as skilled architect and find the funds to deliver what we planned.

In his leadership programs, Louis Tice gave an example from his experience. It seems that one year, close to the end of the school year, Lou's children were acting particularly lackluster. At that point in his career, Lou was a high school teacher. He asked what the problem might be. They responded that, with the end of school in sight, they would not see their friends over the summer as they did at the present, and they were disappointed that the year was coming to a close. Trying to give them a hopeful outlook, Lou responded by saying that perhaps they could put a swimming pool in the backyard and have pool parties over the summer. The children were ecstatic and agreed that it was a good plan.

Later that evening, Lou reflected on what he had agreed to. He had promised a pool in the backyard, but he had no resources to make it happen. What to do?

The next day, when Lou entered his high school at 8:00 AM in the morning, he happened to see his colleague Ed in the cafeteria. Ed was a maintenance worker who often helped the high school find items that fit within its budget.

Lou said "Say Ed. Do you know where I can find a swimming pool? It must be readily available. And I forgot to mention that I have no money for such a pool. Do you have any suggestions?"

Ed thought for a minute and then said, "Come with me."

So, on a Tuesday morning at 8:45 AM, Ed and Lou drove to the end of the runway at SeaTac airport and got out of the car. Lou enquired why they were there. It certainly did not seem obvious to him.

Ed said "Lou, the airport authority is going to lengthen this runway, and they have to move that subdivision at the end to make room. Look out there. You can see any number of above-ground swimming pools. I'll bet you can get one of them for free."

Lou's problem was partially solved. He had relied on his memory to know that Ed was a reliable source and a "go to guy" in a pinch.

Every one of us at some time or other has given his subconscious a problem or a need to solve. Inevitably, the mind works when we don't really know what it is working. Now, this may seem like a very simple example. But try it out next time you have a dilemma regarding resources. It may surprise you.

PROJECT MANAGERS, MAKE BOLD PLANS AHEAD OF ACQUIRING RESOURCES, COMPETENCIES, AND CAPABILITIES

As referenced above, Lou Tice's teaching and principle was that managers do not have to have all the resources at their command to make bold plans. Indeed, it would be rare for them to do so. But many managers act just the opposite. They think they cannot make a move with any plans until they have all the resources at their command.

Lou's theme was to make those bold plans and then find the resources and capabilities needed through the generation of internal energy that the mind provides in seeking a "solution" for the problem. The "reticular activating" system of the brain and mind would take over and uncover all types of resources and solutions.

Here's a small-scale example of this behavior in action. Many years ago when my wife and I had been married a few years, she decided to teach an evening marketing course at a local business college. The course material and the textbook required by the college for the curriculum were very straightforward and followed the theory of business marketing at that time. But it was not very up-to-date in terms of examples and cases which represented present-day marketing and sales practices.

So, armed with the objective of finding good present-day practices, we both began to scour the current literature for examples. Good examples began to literally jump out of the magazines and newspapers about new marketing approaches being employed by individuals and companies alike. At my place of business, I talked with advertising, marketing, and sales managers about new methods of marketing in business. Suddenly we had more material for each class that she could possibly use.

Expressed in other terms, once my wife and I focused on a plan for finding current marketing practices; the gap established between the desired state and the current state produced the internal energy to close the gap between the two.

Instead of trying to assemble all the resources and capabilities to make and carry out bold plans, project managers should define those plans and then search for ways to close the gap between the current state and the desired state. The internal energy and "reticular activating" system will help project managers find the resources and capabilities.

Of course, there are some constraints within a corporate environment that might dictate tempering enthusiasm for bold plans. No project manager would tell his PMO manager that he was going to increase his project team by twenty members and accelerate schedules for projects without a rationale for such a move. The organization might have other plans for those twenty people.

But, without being too literal about the application of Lou Tice's principles, individuals should make bold plans and pursue challenging initiatives. They will indeed find the energy and resources and capabilities to carry out even greater initiatives. Project managers should follow this lead, and the results will follow.

PROJECT MANAGERS: BE CERTAIN THAT YOUR MESSAGE TO YOUR PROJECT TEAM IS THE RIGHT ONE

When I was a project manager with UNOCAL 76 Products Company in Southern California in the mid-1990s, one of my roles was to manage the build-out and opening of convenience stores (C-stores) according to our UNOCAL Fastbreak Program. C-stores were added to some existing 76 products fueling sites, and C-stores were also introduced at new sites. Typical sizes for these C-stores were 1500, 2000, or 2500 square feet of interior space.

The C-store employees who operated and maintained the facilities attended a training program at the Marketing Training Facility in Pasadena, California, prior to the opening of a new C-store. I had the opportunity to attend one of the training programs for new employees.

One of the presenters at the training program was a category management manager who was in charge of working with suppliers to acquire sales products for the C-stores and manage the initial layout of products and floor space. Following the opening of the C-stores, the category manager would assist the store manager in evaluating the best-selling products for the store and making changes in layout as necessary.

I expected the category manager to explain the initial store stocking rationale and to discuss how his team would work with the store manager and his employees to optimize the ongoing stocking for the store.

But as the presentation unfolded, the category manager first emphasized how hard his staff had been working to determine initial store stocking and layouts. He then turned to the suppliers and told them how much his staff had negotiated with the suppliers to make sure the "best" products were identified and stocked. He then turned to some stories about mistakes in initial stockings that had been made and how his staff had coped with those mistakes. He concluded with a "challenge" to the store employees and managers that they were the "front line" and would be evaluated as such.

As I left this presentation, I thought that perhaps I could have been mistaken, but it seemed to me that the category manager had missed his opportunity to train. The feedback I received from some of the store managers was that employees were left in the dark about stocking, layout, and subsequent change process for products.

Of course, there was enough feedback that the category management staff reoriented the presentation to provide the training it had intended to deliver.

But this story raised the question in my mind: **How often do we intend to deliver a message to our project teams and it is really the wrong message?**

The key here is to understand the motivation in the communication. If we focus on our own efforts and not what the team needs, we often miss the point. Learning and training are very general terms, but they must be tailored to the exact message you want to deliver.

So, project managers, be certain of your communications to your project teams before you deliver the message. Understand your motivations. Understand what outcomes and results you want. Make sure of your "audience." Put yourself in the place of the recipient. As such, do you know what to do with the message delivered?

Ken Blanchard, author of *The One Minute Manager* and a noted leadership mentor, introduced "situational leadership" many years ago to assist managers in adjusting their leadership styles to the situation at hand. The communication that goes along with situational leadership is so important to managers. Evaluating the situation is crucial, but delivering the communication that adds value and makes it "actionable" is the key.

It's part of the project manager role. Embrace it. Your words to your project team will live on long after you have moved on.

SAY WHAT YOU MEAN AND MEAN WHAT YOU SAY

The other day I was listening to a speaker on TV talking about his experience with a waiter in a restaurant recently. He was focusing on the expression or phrase "No problem," which the waiter continually used during the process of waiting on the customer. The TV speaker indicated that he was very thirsty when he sat down at the table and initially said to the waiter "Please bring me a glass of ice water with lemon." The waiter's response was "No problem."

Now, the expectation of the customer was that the waiter would bring the water immediately since the customer was very thirsty. The waiter's interpretation of the request was that he could bring the ice water with lemon when he was free from his other IMMEDIATE duties of taking lunch orders, serving customers their food, and circulating among the customers to make sure the lunch went smoothly.

When the customer failed to get his ice water with lemon the next time the waiter appeared at the table, the customer said "Where is my water? I am extremely thirsty today. I need water."

Now, what is happening here? A difference in expectations, along with a lack of clarification of just what each person needed is at play.

How many times have you as a project manager thought to yourself "No problem" when a stakeholder asked for an additional report clarifying how something in the project process was to be handled for that customer? Is there a mismatch in expectations, or are you very clear about when and how the stakeholders' needs will be met? Do you clarify what you expect from the stakeholders as a project manager, because keep in mind that this road runs in both directions.

Think about this. It may mean the difference in everyone being kept current with project plans, results, and expectations. You will be surprised how often this very situation happens in modern PMOs.

WHY MUST PROJECT MANAGERS, ESPECIALLY IN A PMO SETTING, DEVELOP A LIFE-LONG LEARNING ATTITUDE TOWARD THEIR PROFESSION?

When we were growing up, primary, middle, and secondary school were required because the state was assigned the responsibility of preparing children for "grown-up" life. Our personal attitudes toward learning were formed during those years when we were progressing through the grades.

If a student had a bad experience with a teacher or with a subject, often that bad experience would carry over into adulthood as a negative attitude toward "learning" because that student associated learning with that bad experience.

On the other hand, those students who were very successful at navigating their "education" years often became life-long learners.

Now all this may seem very "elementary" to those of you who are engaged in project management today, but, if you really think back upon those school years, and then think about how you approach learning new project management practices and principles today, there is an immediate transference of attitudes from our early school experiences to our adult attitudes toward learning.

So what does this all have to do with whether you seek training and learning for project management practices in your day-to-day work?

Well, it means a great deal.

Those individuals who, during their formal training in project management, developed an attitude that they "learned all that needs to be known," often feel that way because they were turned off by the notion of revisiting grades and tests and unflinching teachers.

But nothing could be farther from the truth here!

Let me "reframe" this perspective (as we say in project management).

Jillian Michaels, the trainer from the TV show "The Biggest Loser," expressed in the May 2010 issue of *Success* magazine article, and its accompanying audio CD, that she believes that people's capability to handle situations depends on how well they understand their own strengths and weaknesses, and how motivated they are to accomplish the task at hand. She says that capability depends, to a large part, on education regarding all factors and issues that impact a person's ability—or lack thereof—to do the job. And since business contexts and interfaces with new people and new situations are arising all the time, it is really necessary to develop a "life-long learning" attitude about your goals and objectives.

Her bottom line is that empowering decisions by people, that move them forward to accomplish their goals and objectives, are the result of their personal education about the issues that make them capable to perform.

Read that sentence again: "Empowering decisions by people, that move them forward to accomplish their goals and objectives (in any walk of life), are the result of their personal education about the issues that make them capable to perform."

Those who continue to view "education" and "learning" in the context of their bad primary, middle, or secondary school experiences are destined to repeat mistakes and fail to prepare to take on life's challenges. Indeed, the challenges facing a project manager in today's modern corporation, especially those PMs working in a PMO environment, are continually changing as new interfaces, stakeholders, scope changes, and cost or economic constraints arise.

Are you a project manager who looks at his own personal training in project management or at training for your team as just another cost to be managed? Is your attitude that training is a waste of time that you don't need to allocate any resources toward? Do you think that training doesn't impact project performance or execution capability? If you are, consider why you believe as you do—did you have a bad experience with "education" or "training" and/or the formal training mechanisms that were forced on you? Or did someone tell you that getting that Mechanical Engineering degree meant you'd never have to study again?

If you have such an attitude, you are shortchanging your project team, your company, and especially YOURSELF!!!! Wake up to the reality that changing circumstances and situations and issues in projects and the PMO require new solutions. These new solutions often may have already been developed by someone else who has encountered and successfully solved the mystery or problem at hand.

Confucius once said:

"By three methods we may learn wisdom:

First, by reflection, which is noblest.
Second, by imitation, which is easiest.
Third, by experience, which is bitterest."

Develop that life-long learning attitude. Embrace everything you can from successful people and their approaches to problems. Immerse yourself in "learning" so that you can draw on the experiences of others. Cast off those attitudes that "training" and "learning" are just another cost to be managed. Or else …

… the bitterest feelings will be your experience.

CAN PROJECT MANAGERS MAKE THEMSELVES "LINCHPINS" IN THEIR ORGANIZATIONS?

Seth Godin's *Linchpin* is a very thought-provoking book about the realities of today's job market and worldwide industry outlook.

In the book, Seth defines a "linchpin" as a person who is "indispensable" to an organization. A "linchpin" is a person who acts as an "artist" by:

(1) creating and sharing new organizational knowledge;
(2) facilitating effective connections among colleagues;
(3) helping develop creative responses to sustain the organization going forward; and
(4) facilitating the organization by dealing with complex and often nonlinear system dynamics issues that, if unaddressed, may hinder sustained growth.

While Seth did not specifically address whether this concept could be successfully applied to project managers, there is no reason why these basic concepts would not apply. The only difference might be the very nature of projects, which are themselves defined by specific timetables, with start and finish dates. In applying the lessons of *Linchpin* to the discipline of project management, it is more effective to focus on the project manager's contributions to the overall corporate environment, rather than his or her contribution to an individual project.

When I read the book, it reminded me of a scenario I experienced about twenty-five years ago, when I had just been promoted to Manager, Manufacturing Performance Reporting and Analysis, in the Controller's staff of ARCO Products in Los Angeles. One day at lunch, the Controller (who was my manager's boss) stopped by my office for a brief chat. We had not had much time together since the promotion, and he wanted to provide his insights as to how I could "be recognized" in the organization.

His advice to me was to do something "extraordinary" that would use my unique talents and would be recognized and applauded by ARCO management. He told a short story about how, a short time earlier, he had himself been recognized. At that time, ARCO was in the process of building a new natural gas-fired cogeneration facility at the ARCO Los Angeles Refinery. The controller was on the finance staff and offered a creative financing idea for the new cogeneration facility based upon his previous work with the financial community. Management had endorsed his initiative, and it had saved the company quite a bit in total financing costs, as well as put the company in a better position to pursue a joint venture with the local electric utility in Southern California.

The controller's advice to me was to look for something similar that would leverage my unique talents in "engineering" and "venture development." You might say it was an early "linchpin" pep talk.

That was extremely good advice. And it is advice that applies equally well to project managers in their day-to-day project work. Look for those contributions that are unique to your talents and which will be recognized for their value to the entire organization, and not just to the single project on which you are focusing at that time. While that may seem like a tall order if you are a project manager who is "up to his rear in alligators" every day, today's market requires that you invest in making yourself "indispensable."

How does that saying go? "The markets may be strong or weak but strong people will endure."

Nurture the four abilities of a linchpin that I summarized from Seth's book in the introductory paragraph. You have the ability to become an artist and to share your gifts with the rest of the organization.

ARE YOU LEVERAGING YOUR CREATIVE SUBCONSCIOUS TO HELP SOLVE YOUR MOST PRESSING PROJECT ISSUES?

In a program entitled "Achieving Your Potential and Investment in Excellence" that I attended when working for the ARCO, leadership guru Louis Tice of The Pacific Institute introduced me to three principles that changed my life and my approach to teaching project planning.

The most significant principle is that, in order to make bold plans, while you will need to ultimately locate the resources needed to carry out the plans, it is not necessary to have all the resources at your command before you make the plan. Your creative subconscious will step in to help you complete the plans.

Your creative subconscious is like a fine ARCHITECT, using your thoughts and intentions and turning them into actions. What is it you intend to accomplish with your project? Like a fine ARCHITECT, your creative subconscious understands your vision and takes over and directs the implementation.

Tice's second principle is that a person moves toward and becomes like that which they think about.

So what was the third principle? People act in accordance with the TRUTH as they believe it to be. This is the reason that people bring different perspectives to a situation.

Believing in one's future by relying on your creative subconscious has many concrete manifestations.

For example, did you ever feel stumped by a pressing problem, only to find the next day that you had some new approaches to the problem that did not seem at all plausible the previous day?

Or, when you get frustrated by a particular problem, you storm off and take a walk, only to find that you come back with a solution?

Or, when it seemed like no one could help you think clearly about solutions, a new contact entered your life who was able to help you break down the problem.

It's all in believing that your creative subconscious is so powerful an architect that it holds answers that you never thought existed.

So, as a project professional, consider your creative subconscious as an integral part of your project team. Give it a role and see how it contributes to your success.

To paraphrase Albert Einstein: in the depths of crisis, you will find opportunity. The next time your project team or PMO faces a crisis, let opportunity prevail. Look to your creative subconscious for a solution.

WAS IT IRRATIONAL DECISION-MAKING OR JUST REFRAMING THE POSITIONS?

One episode of the television show Parenthood has relevant lessons for readers. One of the main characters, Adam Braverman, is faced with a number of issues. He is the father of a teenage daughter and a grade school son and his wife is about to give birth to another child. He has recently lost his job and is evaluating several possible jobs, each of which is really at a lower level than his previous position.

Here are the issues he is faced with in this episode:

1. His son has a learning disability which causes him to speak out in class inappropriately.
2. His daughter is dating a gentleman (Alex) who has a criminal background and is now being charged with assault and battery by the parents of another teen whom Alex fought with at a recent party.
3. His brother is seeking to get a lease on an old recording studio in hopes of reviving the studio. He has used Adam's background information—without Adam's knowledge—to secure a position to bid on the recording studio space.
4. He has no traditional job or income at present.
5. His wife is about to give birth, and there is a possibility that the baby could also have the same learning disorder that their son experiences.

Now, we all talk a lot about rational decision-making being the basis for business interactions and business contracts.

Adam thinks that the rational choice is for him to accept a job that provides an income, but not a fulfilling experience. He also knows that there is very little he can do about the assault and battery charge against his daughter's boyfriend, but, in his mind, it's just another indication that their relationship was a mistake from the beginning.

So, what decisions does Adam make in this episode?

First, when Adam and his brother meet with the landlord of the recording studio building, they are initially met with rejection because the landlord says that he wants a safe and reliable tenant that he can count on. At the heart of this conversation, Adam takes over the dialogue and tells the landlord that he believes in what his brother is doing. What's more, he tells the landlord that he has business skills that he has not tried out in this context before, but which could be exactly what the new startup recording studio needed. In doing so, Adam surprises both the brother and the landlord.

Second, when Adam's daughter tells him that she feels really powerless because she can't help her boyfriend with the charges leveled against him, Adam tells his daughter that everything will work out. He then goes to the parents of the assaulted teenager and speaks to them from the heart about how he first doubted Alex but was impressed when he learned how Alex had moved from using a soup kitchen as a source of food to actually running the soup kitchen. Adam's eloquence in this dialogue resulted in the parents dropping the assault charges leveled against Alex.

Now, was this irrational decision-making? At first glance, it appeared to be just that in the face of harsh facts that pointed in one direction, he reversed his field and stood up for his brother and for Alex.

Meanwhile, the side story about Adam's son's learning disability, and his inability to get along with the other kids in the class, developed into an asset. When the kids in the class found out that Adam's son had advanced to a high level in their favorite video game and knew how to beat a major obstacle, the kids sought him out as an authority who could help them achieve the same success with the game. His influence with the classmates increased. Seizing upon this success, Adam's wife and the teacher developed a strategy for Adam's son that they hope will result in him getting along better with his classmates.

At the end of the episode, Adam is still faced with no income and an uncertain position in his brother's recording studio venture. He feels, however, that he "did the right thing."

The real story here is how reframing the issues provided Adam with a greater ability to make difficult decisions. Until faced with a decision, Adam had not focused on his own skills and competencies to see how they might apply to his brother's recording studio venture. Adam was uncertain about how to proceed with Alex until they learned more about his past, and how his drive to succeed and survive led him to run the soup kitchen.

Reframing is a very important tool in project management and in life. It is why I tell my Project Lessons Learned to students to get as many perspectives as possible when describing a "Significant Event" that may eventually become a "Candidate" for a Lesson Learned—"People act in accordance with the truth as they perceive it to be." Everyone sees different outcomes from the same scenario.

In this episode of Parenthood, Adam reframed the issues and acted at the right time, with the right intent, and with the right IMPACT to make a difference.

Every project manager and every PMO leader is, at some point, faced with a difficult decision. It is our job to face the facts, reframe if necessary, move toward action, speak with conviction, and lead by leading.

STORIES ARE THE ESSENCE OF LEARNING

A popular book from many years ago has resurfaced. It is *The Five Love Languages* by Gary Chapman. It is an excellent book about relationships and the interactions between humans that sustain and enhance relationships. In the five chapters that introduce the five languages, every chapter begins with a story. The story tells the relationships between two or three participants and, in every case, the outcome reveals the thesis of each chapter.

I bet that after readers put the book down and try to recall the titles of the five languages, they will be in a better position to relate the details of the story, than in reciting the exact titles.

Stories are the essence of learning. You will learn more from the stories that make up your human experience and the stories you hear from close associates whom you respect than you will learn from all the textbooks that have ever been written.

Now, that is a bold statement, but it really is true. That is why I have always paid attention to stories that subject matter experts, project managers, and leaders continually tell in their training and leadership development classes. Stories provide the human elements of situations and scenarios.

Stories are the essence of our lives and our experiences. Oprah Winfrey was interviewing President Jimmy Carter in his Plains, Georgia, home once and asked the former president how he liked to begin his day. Carter remarked that like most southerners, he enjoyed a cold glass of orange juice to begin his day. Then he paused and said, "[a]nd of course, I bring a steaming cup of coffee to Rosalyn." I think this story is why I like to begin my breakfast each day with a cold glass of orange juice.

Regis Philbin's book *How I Got This Way* tells many stories of Philbin's life and its relationships with the broadcast media. One story he relates is from his student days at Notre Dame. The football team at that time was considered invincible, and it was expected that they would win every game. But when they lost a crucial game to Purdue, the student body and fans were beside themselves to explain what happened. Coach Frank Leahy appeared after the game outside the Notre Dame locker room and addressed the fans. He stated that, in life, there are often setbacks that can't really be explained and that assigning blame for the result only dilutes the entire effort. He said the team would persevere and be a winner again. His story was just the right medicine that the students and fans needed. In another situation, Philbin was faced with rejection for a Chicago morning TV job he expected to get. As he was driving on a Chicago freeway toward home in a dejected state, he suddenly saw the exit that would lead him to South Bend. Instead of going home, he drove to South Bend and was counseled by Coach Ara Parseghian who once again told him that rejection wasn't the end of the world. Philbin even had the chance to play catch with famed quarterback Joe Montana on that visit. These were tremendous stories that guided Regis Philbin throughout his life.

In my book *Project Lessons Learned: A Continuous Process Improvement Framework*, I relate a number of stories to emphasize points and lessons I really want the reader to remember. One is the story of a project manager during the merger of Conoco and Phillips who was assigned the task of integrating the systems of four geographic groups into one group and his method of communicating to everyone his plan and their role in making it happen. He actually traveled to all the locations and talked to everyone. It fostered a commitment to his plan and made it successful. This type of empathic communication is to be valued over the manager who only talks to the top management in putting together and initiating his integration plan. The latter situation, in my experience, never fostered great collaboration or commitment to the end goal, and it seems to have been the norm for many mergers and integration activities for many years in the past.

My message to project managers who want to build successful project teams is to rely on stories to help their teams understand how other project teams have behaved and to capture their own experiences in current projects and relate them to future project managers who might benefit from their experiences. In my experiences in a number of PMOs, I have always valued the stories that are told and retold about situations that worked well and others that turned out with disastrous results. These are really the essence of the PMO experience and lessons learned. Project managers should listen to their project team members and groups to learn what is really happening in areas of the project where visibility is often hidden by the day-to-day weeds of activity. When project teams share their experiences, it fosters a better commitment to the overall goals of the project and growth for the entire PMO.

OKAY ALL YOU PROJECT MANAGERS—LET'S REFRAME AN ISSUE

So far, this book has talked a lot about how you can take leadership or a problem by reframing it to find a solution. When and how does a reframing take place?

Let's take a real-life example to guide our thinking here. This example is from a Major League Baseball playoff game. At one point, the New York Yankees were playing another team, and Mark Teixeira, the first baseman for New York, was on first base. Mark had suffered a calf injury late in the season that impacted his running, but he had returned to the team just before the playoffs. The announcers for the game speculated whether Mark would try to steal second base in this game. They concluded that the calf injury was such that Mark would be content to stay on first and let the batter hit the ball to advance him to second base. Prior to the next pitch, the first baseman was not holding Mark close to the bag, so Mark took a big lead. When the pitcher came home with the pitch, Mark took off for second base and easily stole the base standing up. The announcers looked at each other puzzlingly because the action was contrary to their expressed thinking on the broadcast.

But a "reframing" of this situation in the game revealed that Mark's action was just the action called for in this situation. First, the Yankees had not been hitting the ball well, or scoring runs in the playoff series. Second, the first baseman, in not holding Mark on first base, was really allowing him to take a larger-than-normal lead toward second. Third, the Yankees needed a run, and something to jump-start their momentum. When the announcers reflected on these additional facts about the situation, it seemed a natural thing for Mark to steal second base. As it turned out, the batter got a base hit to the opposite field, and Mark scored a run from second base on the hit. The stolen base was really a smart play. The likelihood of him being thrown out at second was negated by the huge lead the first baseman allowed. The surprise move was just what the Yankees needed to get their scoring started in that game.

Sometimes, in situations like this, we are shortsighted in our evaluation of all the relevant facts in the situation. As a result, our decisions are based on limited inputs that we have chosen to use based on the defined situation. However, as in the case of Mark Teixeira and the stolen base, a more complete look at the relevant facts to the situation revealed some options we might not have thought of.

In your project work, how often do you take a second look at the relevant facts in a situation before making a decision and proceeding with the action defined? Perhaps the right thing to do would be to get some additional input from those around you who see the same project but perhaps with a different "perspective."

That's what I am advocating for you right now. Take a close look at all the relevant facts before making that decision. It is not always possible to see all the relevant facts in the situation, or they may be obscured by the actions of others around you. But "reframing" remains a powerful way to present to yourself or to others options in strategy that may lead to very successful outcomes. It's your call.

FOR ALL COMPETENCIES AND CAPABILITIES, THE LITTLE THINGS COUNT JUST AS MUCH AS THE BIG ONES

I once worked for a manager in a project office for a major corporation that assigned various responsibilities to his project staff members so that our group objectives of being the benchmark organization for both project management and process management could be assured on an ongoing basis.

As an organization, we had just moved into a new building. At that time, conference rooms were not installed with modern video and acoustic equipment to facilitate conferences or presentations. The audio-visual equipment was maintained by the HR administration group, and every time we needed to hold a meeting, a staff member would need to call HR administration and reserve the appropriate equipment for the conference room. On the date and time for the meeting, the staff member was responsible for securing the equipment from HR and setting it up in the conference room. At the conclusion, the reverse order was followed. Ultimately, what this meant was that a staff member could never be sure whether the equipment were, in fact, on hand when needed, and often last-minute changes occurred to accommodate missing equipment.

As you might have guessed, my manager assigned me as the AV coordinator for our group. My first reaction upon hearing about this assignment was to be "miffed" that I would have to handle this assignment, along with my usual roles as project manager, standards developer, benchmark specialist, and process authority. What did I do to deserve this lowly assignment?

If you are familiar with Herb Cohen's screen stories about negotiations, you will recall a story he told once about his wife actually buying a home on the North Shore in Chicago by getting all the stakeholders around her to agree that it was a good idea. Herb described himself as the "technical titular leader" of the household in making such decisions, and in negotiating the terms, but his wife had actually taken a lesson from his book and surrounded him with the ultimate stakeholders who would ratify the decision. He was rendered no longer the "technical titular leader." Such is the way I felt about my new AV coordinator assignment.

As I agonized with this assignment for the group for about a year, I began to develop some insights about what good AV equipment and installation should really be. I became an expert on bulb change-out, electrical failures, projection screen mishaps, and general "where to get it done in an emergency."

To make a long story short, I actually began to appreciate the role. As our group's needs expanded to be able to connect remotely with AV sources and to enhance the effectiveness of our presentations, I began to realize that this AV competency was just as important as the project management methodology I had been wrestling with as I interfaced with our major corporate consulting organization.

Over time, I began to realize what an important role the AV coordinator was to the Project Office, and to the corporation. It was the entranceway for knowledge to flow into our groups and into our heads. Well, maybe not exactly, but you know what I mean.

The message here ought to be very clear indeed—it's the little things that make up the larger competencies and capabilities that are so important to the overall success of a role or a career. Often the messages that dominate the little things are the same messages that dominate the actions and decisions of an individual or of a group. But it's often hard to see for yourself by looking at it through your own eyes. Getting out of your own body, it would be so easy to see how every little thing you learn contributes to your being able to do almost anything in life. Keep this in mind the next time someone asks you to take on an assignment that you think is beneath your standing or is not worthy of what you aspire to be. You may learn just what you don't know that will lead you and others around you to success.

SO HOW DO I SPECIFICALLY APPLY THE SMART GOALS FRAMEWORK IN THE CONTEXT OF PROJECT MANAGER GOALS AND OBJECTIVES?

Choose five project deliverables for your project—these are developed by groups within your business. You have little control today over the work plans and priorities for developing the project deliverables.

Develop a plan for the five project deliverables which outlines the following:

1. How do you plan to influence the groups in developing the deliverables on time and on budget?
2. How do you plan to track progress on the plan in (1) above?
3. How do you plan to mitigate the risk involved if the deliverable appears to be late or not to budget standards?
4. How do you plan to interact with other groups who interface with the five project deliverable groups to ensure that they contribute any portion of the deliverables to the primary group?

THE REAL POWER OF YOUR NETWORK

Do you think that you can snap your fingers whenever you need some vital information?

Do you think that your LinkedIn connections or your Facebook or Twitter friends will come running to your aid whenever you need help?

If it was late at night, and you were sitting in front of your laptop, in a foreign country, several thousand miles from most of your network, with a useless cell phone because your provider had no network, and you needed help, what would you do?

When this happened to me on a consulting engagement, I was thrilled by the responsiveness of my network, and I just had to share this fantastic networking experience.

When I was in Panama facilitating two three-day courses on Project Lessons Learned for the Panama Canal Authority's Construction Division, at the end of the second day of the second session, the division's manager requested that the next day I speak to the class about some specific Project Lessons Learned success factors and barriers experienced by other organizations. He also wanted to know how these success factors and barriers fit into a cohesive Project Lessons Learned system. He requested this information because he wanted to impress upon the class that Project (and Contract) Lessons Learned would (and should) be an ongoing part of their daily activity.

I left the building that day worried because, while I had many examples, anecdotes, and ideas about Project Lessons Learned success factors and barriers, I was not aware of any single document or study that would address his request. Since I unfortunately hadn't anticipated his request, after dinner that evening, I sat myself down in front of my laptop and began to brainstorm about who in my network might be able to efficiently guide me in the right direction—time was of the essence since I had to be back in the classroom about eight hours later!

During my brainstorming, I remembered that Michael Guidry's "Northwest Arkansas PMI" chapter recently featured Deborah Grassi, a senior manager in SAP Change, Training, and Communication at WalMart. Deborah spoke about lessons learned from a significant WalMart project. So, I dashed off a quick email to Michael explaining my conundrum.

I sent another quick email to Dan Ranta, a director of Knowledge Sharing with whom I had worked at ConocoPhillips.

I also recalled that Lisa Austin, a manager of Knowledge Management at Williams Midstream, had recently spoken to the PMI Tulsa Chapter about Knowledge Management at Williams. While I had unfortunately missed her presentation, I thought contacting her was worth a shot since she and I are connected on LinkedIn. Lisa quickly replied that, if I had not already seen the website and blog on Knowledge Management in the United Kingdom operated by Nick Milton, director of Knoco, Ltd., I should check it out.

Nick Milton was formerly a Knowledge Management executive with BP. Now, he is a Knowledge Management consultant who consults with corporations worldwide on Knowledge Management issues.

I quickly accessed Nick's website and found a section on Lessons Learned that included a survey that he had conducted in 2009 of companies in many different industries. The survey discussed these companies' varying Lessons Learned practices, which factors had been barriers in establishing Lessons Learned, and which factors had led to success in Lessons Learned and lasting change within the organization.

BINGO!!!

I immediately messaged Nick on LinkedIn and introduced myself. I explained that we had common interests in Project Lessons Learned, that I was in the midst of teaching a course on the subject, that I had been referred to his website, and that I was very impressed by the survey. I requested that we connect on LinkedIn. Almost immediately, Nick approved by connection request and explained that he was with a client in China but that he would love to talk about our common interests in Project Lessons Learned.

Nick and I emailed a few more times that evening, and Nick was kind enough to permit me to present his survey and discuss its findings with the course participants.

The survey had seventy-four responses from individuals in such varied industries as oil and gas, engineering and construction, consulting, mining, industrial products and services. It mainly focused on project groups, and it covered success factors, barriers, key systems components, and other enabling factors.

As I read through the survey, I was pleased that it supported many of my recommendations, including the use of the After Action Review (AAR) as the preferred basis for a Project Lessons Learned framework, as well as the inclusion of risk management when identifying candidates for Lessons Learned.

The next day, I distributed the survey to the course participants, and we discussed its findings within the context of the Construction Division's business and project or contract environment.

Throughout the day, similar helpful information flooded in from my network. In fact, nearly one hundred percent of the people that I contacted responded with useful information.

I learned **two valuable lessons** from this experience: (1) when training, listen—and be responsive—to your audience, as it often provides valuable insight as to how you can develop and improve your materials; and (2) believe in the power of your network!

Remember: the responsiveness of your network is entirely up to you.

As Robert Cialdini has stated in his work on **INFLUENCE, RECIPROCITY** is a powerful motivator. Always do what you can for others before asking a significant favor in return.

Your network will only be as responsive to you, as you have been in fulfilling its needs.

PROJECT MANAGERS, IT'S TIME TO TOOT YOUR OWN HORN

Why do golfers watch "The Golf Channel" on TV? I believe that at least 90% of the viewers already know how to play the game. Many of them are so good that they compete in tournaments all over the world. But they still watch "The Golf Channel" every time they get a chance.

Well, the truth be known, it's probably because all these golfers want to get better at playing the game. They are looking for those little tips from the experts, the PGA and LPGA pros and legends of golf to guide them to new ways to play the game. They are looking for those "lessons" from the golfers who make golf their livelihood.

And, of course, every golfer defines for himself what it means to "get better." All are looking for outcomes with their golf games that satisfy some inner objective they have in their lives.

Now if you examine the topics and approaches of many of these "lessons" programs, they center on helping golfers change something about their game that is hindering them from scoring like they really want to score.

But, if you look closely, they are also taking another approach these days. Some golfers have developed what I would term "Best Practices" in their swing or their putting or their short game that continue to work for them day in and day out. It's termed "success." Some of the legends of golf have those attributes as well. If you look at video footage of Arnold Palmer or Jack Nicklaus, they each have a unique mannerism in the way they swing the club. And it continued to serve them well throughout their careers. So one of the things "The Golf Channel" is focusing on right now is the uniqueness that each golfer brings to his own game.

Everyone around you is always telling you to do it a different way which follows some group's established project practices with regard to projects. And your PMO manager is telling you to watch that budget and don't let that schedule slip. But there are some things that are unique about what you do every day as a project manager. They make you what you are in your particular project group and business context.

Embrace that uniqueness … let it guide you to new heights of accomplishment in project management. You will be a guide to others and set new standards for your project's success.

A TALE OF TWO CS—HOW PROJECTS AFFECT LIVES

Sometimes project managers are as much informed by their projects, as they inform their projects. And sometimes the lessons they learn are as valuable to the broader scheme of life, as they are to their projects going forward.

What does that mean?

It means that the total experience of carrying out and managing a project sometimes adds richness to a project manager's experience that no training, no development course, no experiential modeling, no simulation, and no textbook could ever provide.

It is these experiences that project managers should reflect upon in order to draw out some real lessons when dealing with the physical, mental, social, and emotional states of project management.

This is the story of one such encounter from my experience. I hope it provides you with some insights, and I hope that you share your own similar experiences in projects so that others in the project community may benefit.

In 1996, while working for UNOCAL 76 Products Company as a Marketing Operations Analysis project manager, I was pressed into service opening convenience stores under our "Fastbreak Format" program. This was an initiative to construct and operate convenience stores ranging from 1500 square feet in interior size to 2500 square feet interior size. Many of these convenience stores also included fast food sections that featured Carl's Jr., Subway, or another food brand.

As a project manager, it was my job to take the site from the construction group and then implement a plan for the completion of the store's interior layout (including fixtures for the display of soft drinks, candies, and snacks) while also completing the store's exterior layout (including underground fuel storage tanks, fuel dispensers at the islands, and other features). Systems in the store included Point of Sale equipment that was sourced by major suppliers and installed by a special UNOCAL systems group.

Obviously, coordination and timing were of the essence in such an operation. I had at least ten bosses from various groups who always wanted to make sure that their portion of the exercises was completed on time, on budget, and fully operational.

There were two C-stores in particular that I remember from this experience.

One was located in an inner city area of Los Angeles. It was a 1500-square-foot interior C-store that was to be franchised by a couple in the local area.

The other was a 2500 square foot interior C-store that was to be a company-operated store on a freeway location just north of San Diego.

The larger C-store was equipped with several islands for gasoline dispensing, while the smaller C-store had only two islands for gasoline dispensing due to the small lot at the inner city location. The smaller C-store was equipped with a bulletproof glass partition that separated the operator from the customers after certain hours of the evening. The larger C-store, on the other hand, had an open and spacious layout with extensive coolers for a variety of cool drinks and beverages.

The larger C-store had a company-employed manager who was already a manager at a nearby C-store about ten miles from the new location. Although it was his responsibility to manage the store upon opening and to assist me prior to opening, it was solely my responsibility to manage the project, interface with all the suppliers and vendors, work with all the company systems groups for the installation of systems, and determine when the store was ready for operation.

In the case of the smaller, franchised C-store, my principal responsibility as project manager was to complete the project, interface effectively with the couple who would operate the store upon opening, and determine when the store was ready for opening.

The couple who had franchised the store had two children under age ten and a grandmother who was always with them during the construction, installation of equipment, training, and opening of the store. Although the manager of the larger C-store was extremely helpful in carrying out routine tasks in getting the store ready to open, he was also very impatient to "get his hands on the reins" and drive the stagecoach.

These two stores were going up concurrently, and I lived about half way between the two stores. I divided my time each week between each location as I managed the various project tasks that had to be staged to bring each store to a ready state.

Occasionally during the work, a representative from another UNOCAL group would appear on-site to talk to either the franchisee or the company store manager. These discussions, of course, always involved me as a principal participant in the preparation of the site.

There were always issues in the opening process that required my liaison with another UNOCAL or supplier group. For example, we encountered systems problems with the installation of the Point of Sale equipment in the smaller C-store because the satellite communication equipment for credit card authorizations was not working properly. So, a supplier from the network systems vendor was on-site much of the time during that phase of the setup, diagnosing and troubleshooting the satellite communications. This meant that I had to describe the problem, the planned solution, and the timeline for fixing it to the franchisees. Although nothing in the project plans called for informing them about how this interface with going, I felt an obligation to let them know exactly what problems we were encountering since they seemed to be at the site twenty-four hours a day.

As the two stores slowly took shape, my drive from my home to each location each day provided a good time to reflect and plan and digest and understand the C-store operators' different agendas. My obligation was to provide an open C-store, free of any systems issues, and ready for steady-state operation.

The smaller C-store was ready first even though we had been delayed by systems problems several times during the project. I began to think of the training that the franchisee couple had received as the store neared completion and wondered if they were really prepared for what was to come. They would have control over when the whole C-store would be open to the public and when customers could interface with the manager only through the bulletproof glass partition. This was in stark contrast to the freeway location for the larger C-store where hours were almost unrestricted. Although twenty-four-hour operation was not common at that point in time, the store was open from 6:00 AM to 10:00 PM.

On the last night when I drove away from the smaller C-store, I was assured that the couple had sufficient training to see the store's opening through. I made sure that company personnel and vendors or suppliers had called on them enough that they had phone numbers and contact information with backup numbers if necessary. The wife asked, "Does this mean we will not see you again?"

I wasn't sure how to react. All I could say was "Yes, I had completed my project and felt they were prepared to operate the store successfully." But that is when I realized that this was more than just a project; it was their livelihood, and lives were dedicated to making this venture work.

As I drove away that evening, I questioned whether I had done my best to prepare them for the days ahead. Had I coached them enough on the credit card system? Had I helped them enough to understand the dispensers and the card readers? Had I really treated them like they were going to be successful in this endeavor? Only time would tell.

Meanwhile, the larger C-store was nearing completion. More corporate people were appearing every day to question the store manager and his staff on their particular areas of influence and control. As we neared completion, I participated in the training of the staff to handle every aspect of the store.

It was solely my call whether the store was ready for operation. As project manager for this C-store opening project, I weighed all the evidence and decided that we needed one more day to be sure that everything was in working order. My store manager disagreed and voiced his disapproval to the Business Development Manager (BDM) for the region. Although the BDM was concerned, he recognized that I was the only person who had been on-site every day, knew every piece of ground, had talked with every vendor, had questioned each systems development specialist, and had operated each dispenser and each card reader.

As it turned out, we did have a minor systems problem on that last day before officially opening—a problem that no one, not even me, could have foreseen. It took several hours to resolve, but I knew better at that point than to make a big issue of why we needed one more day.

As we opened the larger C-store, I thought of the smaller C-store and the vast differences between the two: the differences in location, market, operations, and especially the challenges for each. Had I done my best to prepare each one to succeed? I learned so much from the experience that I am sure it affected my approach to future projects and the people involved.

THREE STEPS TO HAPPINESS FOR PROJECT MANAGERS

Ultimately, after all is said and done, I believe that we all are searching for happiness in our lives. It's that combination of emotions, activities, and long-lasting collaborations that bring us satisfaction and enjoyment from living every day.

When my daughter was married, I gave a toast in which I addressed "happiness." I said that happiness really consisted of three different things.

First, if you think back to the past to those things that you have shared with your loved ones, which have made you feel good and fulfilled, that was a first step to happiness.

Second, if you think about the present day and those events and emotions you share with your loved ones that bring you special joy in the present moment, that is a second ingredient to happiness.

Third, and most importantly, if you think about the anticipation of those things you will share with your loved ones in the future and how you can work to enrich those experiences for each other, then that is the ultimate ingredient for happiness.

Lou Tice has often said, "People move toward and become like that which they think about."

Thinking about and anticipating happy experiences in the future will help to bring them about and make them an integral part of your life.

My toast was very sincere and heartfelt. I wanted to leave a lasting impression for my daughter and her husband-to-be to contemplate as they approached the future together.

Much the same thing can be said about project managers. I believe that each project manager is seeking happiness in his or her own context of the project and the project community. To really achieve "happiness," a project manager must look back, examine the present, and anticipate the future.

As a project manager, where are you seeking happiness in your projects? Perhaps you never thought about a project or a project team being a source of happiness. But, if you believe my initial premise that all of us are ultimately seeking happiness in our everyday lives, then a project manager must really address his entire experience in and out of projects.

4

PMO and Project Performance

WHAT DO I DO IF MY PMO CAN'T EXECUTE?

A number of years ago, the Los Angeles Dodgers and the Cincinnati Reds played a baseball game in which the Dodgers did nothing right and the Reds did nothing wrong. The final score was 10-0 in favor of the Reds. In a post-game press conference, the Dodger manager Tommy Lasorda was asked this question by a reporter, "How do you feel about your team's execution today?" Tommy thought a minute and then replied, "I am highly in favor of it."

Execution. Getting things done–consistent, coordinated, controlled execution. It's the road to success in today's competitive global marketplace. But how do you optimize that execution? How do you correct your PMO's execution when you're not getting the expected results?

I would like to provide one perspective to this question based on my experience working in a PMO and observing the changes in its performance and the actions that we undertook to optimize that performance.

As evidenced in the baseball game mentioned above, execution includes "actions" and "behaviors" and "decisions" on the part of both teams that lead to "results." In a sense, we all develop or forecast "expectations" for those results before they occur and then we measure the difference between the actual results and the expected results.

That gap can be very elusive because it contains not only team or group performance items but also individual performance items. It includes day-to-day decisions and judgments based on information we gather at or near the time of the actions. It also brings the business context into play.

 DOI: 10.1201/9781003593645-4

So optimizing the results from the execution means understanding the "gap" between expectation and actual and then providing feedback to change the execution. Sounds like a process lurking in there somewhere!!

Let's look at an actual example. Some years ago, I was a member of a PMO following the merger of two major corporations. I had participated in the development of the processes, standards, and procedures for the PMO, so I understood the structure of the PMO quite well.

During the first year of actual operation following the merger, this PMO was engaged mainly in systems integration and applications/systems rationalization projects as the two merged organizations combined their business systems and in most cases developed a single system going forward where each had had a system before. An example would be the credit card networks of each company being merged into a single credit card system as a result of a PMO project.

The PMO delivered a number of projects during that first year so that, on a graph of a number of projects completed versus time, the upward trend was along about a 45-degree angle. That performance defined the "expectation" for the PMO. Nobody stated that expectation as such, but everyone had the idea that the project teams were performing such that they could predict over time how many projects would be completed in the next incremental time frame.

After about one year in this mode, the emphasis shifted over to the PMO conducting projects for the business functional groups in the newly merged organization. During the first year of the merged corporation, those groups had been forming and organizing and not focused on systems projects addressing business needs. Over the next year of operation, the curve of projects completed versus time began to flatten out and tended toward some asymptotic level. The PMO was no longer completing projects at a 45-degree clip.

What had happened?

A number of us formed a small group to study the structure and execution of the PMO at that point. We were very aware that there had been a shift from mainly systems integration to new business functional group projects.

What we found was that there were two basic reasons why projects were lagging in execution.

First, the PMO had limited financial consulting resources for consulting with the business functional groups on their business case economic analysis. Such financial consulting was a basic requirement for a complete business case and subsequent funding approval of the project.

Second, we found that the PMO project coordinators and analysts were having difficulty translating the business functional area business requirements into project scope and business requirements for the projects. In other words, we were having difficulty "framing" the projects correctly to answer the question "What problem am I trying to solve with this project?" Subsequently, there were a number of false starts and "reframing" of the projects which led to fewer total projects completed over time versus those in the systems integration phase of work.

Now the first finding was easily addressed. The PMO added additional resources to assist with the financial analysis and in many cases partnered with financial services to coop those consulting resources.

Solving the second condition was a little more complex. When we examined it more closely, we found that the PMO project coordinators and analysts were not holding effective dialogues with the business functional groups to define the project business requirements. We immediately enlisted the help of "Crucial Conversations" training coordinators to teach effective dialogue within the PMO and business functional groups. Very rapidly, we began to "frame" the projects correctly the first time leading to good and timely execution.

So, from a prescriptive point of view, what would I suggest you do if your PMO is not executing as expected?

First, look for alignment issues. Stephen Covey once said, "Every organization is perfectly aligned to get the results it gets" (see Stephen Covey, *Principle Centered Leadership*). Alignment means that business processes, jobs, people, systems, rewards, values, and beliefs are aligned so that they support each other in the accomplishment of goals and objectives.

In the case in which business processes are changed, the other alignment variables must be addressed in order for results to be consistent with expectations. For example, if a business process is changed which involves a change in jobs and people, then systems and rewards as well as values and beliefs must also be addressed. Too often in the past, companies or smaller organizations within companies have changed business processes and the associated jobs and people without really addressing the systems, rewards, values, and beliefs that must support those process changes (see Michael Hammer's *Beyond Reengineering* for the business model). Also, look at the Organizational Effectiveness (OE) work by Booz & Company on this subject.

Second, look for changes in objectives or goals or direction. In the example, the PMO shifted its emphasis from systems integration projects to business functional group projects, which meant a new set of competencies might be necessary to continue with execution as it was expected.

Third, apply some systems thinking to the situation. (Here, systems thinking can be thought of as in the context of Peter Senge's *Fifth Discipline* or Daniel Kim's work.)

In systems thinking, the organization can be looked upon as a system which refers to organizational structure, culture, processes, procedures, feedback, human interactions, policies, and the structure we use to manage the process.

In systems thinking, we can see organizational processes as either reinforcing processes or as feedback processes. Reinforcing processes are those in which more is gained as we push harder on the variable defining the process. Balancing processes are those that tend to retard reinforcing processes.

As a general rule in systems thinking, if a growth scenario appears to be reaching a "limits to growth" situation, then look for the balancing loop that is tending to retard to the growth of the reinforcing loop. Rather than continuing to push harder on the reinforcing loop, solutions can be more easily found by examining the balancing loop.

That is why solutions to the PMO scenario in which the project completions were slowing were to be found in the balancing loop which contained the financial/economic analysis as it contributed to the business case and the lack of effective dialogue promoting proper "framing" of the problem and poor business requirements definition.

Fourth, make sure that any major initiative or project which the PMO pursues has a robust Organizational Change Management (OCM) plan to ensure that all the stakeholders of the initiative or the project are addressed according to what is expected of them after the initiative or project is implemented. If you see results different from expectations following a major internal PMO initiative, look at the associated OCM plan to see how well it performed.

There may certainly be other perspectives on analyzing the execution of the PMO, and I invite others to comment on specific situations they have been involved in with PMO activity. We can all recall a myriad of interventions we have pursued within organizations to realign expectations and results. It is a never-ending job to continually examine alignment, but it is the only method for ensuring lasting, consistent EXECUTION.

HOW TO CREATE A STRATEGIC PARTNER WITHOUT REALLY TRYING

The behaviors of PMO project managers, team members, stakeholders, and others in the project community discussed in this book are real behaviors that I have seen in my experiences working in and setting up several PMOs.

Much has been written in recent years about strategic partnerships in project management. The idea is for parties to collaborate to ensure the long-term success of an organization such as a PMO as projects are executed in support of strategic directions.

Consider the following scenario: A PMO initiates a systems project in support of a major business/functional group in the organization known as the Commercial Group. The Commercial Group is engaged in activities ancillary to the firm's core businesses, and in order to achieve its returns, it operates in markets focusing on short-term, day-to-day operations. The PMO systems project is designed to use a vendor-developed application, and its supporting systems, to provide the Commercial Group's day-to-day operational and transactional focus. The Commercial Group identifies a number of potential vendors and evaluates which vendor could supply the systems requirements on an ongoing basis. It then selects a vendor and development work proceeds until the system is in production.

Over the next several years, the Commercial Group's activities and returns flourish, and the vendor makes a number of requested improvements. Every time a new and updated system and application requirement was identified by the Commercial Group in response to the market, the vendor delivered.

Switch gears for a moment …

A number of years ago while working for Atlantic Richfield Company (ARCO) in California, I had the opportunity to serve on the Board of Advisors for the University of California–Irvine Science Education Advisory Board. The board advised the professors and administrators in the School of Physical Sciences on their K-12 development programs for students and teachers. Funding for the work was provided through contributions from local corporations to the board. Each year the professors and administrators solicited the corporations for funding based on the premise that the K-12 programs would remain the same quality and scope as in previous years. After a few years, it was almost a foregone conclusion that the budget would be the same or perhaps increased slightly from the previous year because additional corporate contributors were identified. An annual report of the board to the contributing corporations provided details of the K-12 programs and the breadth of participation among K-12 teachers and students.

After serving on the board for several years, I was summoned one day by two professors who were instrumental as leaders of the program. Since I had a background in strategic planning and analysis, they asked if I would take on a small project in support of the board activities. They were concerned that funding for the program seemed to be on the decline. Over the past three years, the funding had decreased by $20,000 each year so the quality of the programs was in jeopardy. They were at a loss to explain why the sudden decrease in funding over the three-year period and asked if I would undertake some analysis to assist in their planning.

I was glad to assist in this effort and studied the scenario from various perspectives for a month or so. I interviewed corporate contributors, the teachers involved in the K-12 programs, the professors, and other community groups. I looked at other university groups who approached their funding in a similar way and others who approached funding from alternative means.

Here were my findings:

1. The business climate in California had changed over the last three years. Whereas aerospace and petroleum companies dominated the corporate giving scene three years earlier, biotech and medical device companies were beginning to emerge to take their place.
2. An individual corporate contributor had many more choices in making contributions because many other groups in that California community were doing similar work with schools.
3. Other university groups had emerged as new curricula developed, and these groups increased the number of total university groups seeking corporate contributions.
4. The methods of solicitation of corporate funding were changing. Many groups partnered with the stakeholders in the K-12 schools who received the benefits of the programs to provide a brochure to corporate contributors that tied real benefits to both the provider and the receiver. This method was unlike the solicitation method of the School of Physical Sciences, which relied on its past work to carry the story.

The real themes from this analysis were:

1. Don't rely on "static" data to tell a "dynamic" story. Look at the dynamics of the situation rather than a point in time.
2. Don't consider yourself to be the "Center of the Universe" when it comes to activity. Look at the corporate contributors as the "Center of the Universe" and map the interactions they have with the groups requesting funding. This tells two things. There are more requests than in previous years, and many of the requests tell a story that clearly ties in the end result with the level of contribution.

So, the real story of the fall-off in corporate contributions was in the dynamics of the business climate and the solicitation model. Never look statically when you should be looking dynamically. Never consider yourself to be the "Center of the Universe" around which other activities revolve. It is a recipe for disaster.

Now, what does this have to do with our scenario of the PMO group and the system's application supporting the Commercial Group? Over time, as the Commercial Group grew and its business grew, more and more was asked of the vendor who supplied and supported the system's application which facilitated the transactions enabling the Commercial Group to flourish.

But, there came a time when a new business requirement relayed to the vendor was met with the response "We don't have any more resources to do this enhancement. All our resources are fully employed to support the day-to-day operations."

The PMO was incredulous. It had created a strategic partner without really trying. But it had failed to plan for the development of its partner to support new development work. Why? The PMO focused on the static scenario, rather than on the dynamic scenario. They felt they were the "Center of the Universe" when it came to that vendor's support. In actuality, the vendor's business had grown to support other major companies as well.

This is a good example of how companies can put strategic vendors in place without really trying. It sounds rather farfetched. But it really happened to a major PMO organization. Is your PMO ready to define the capabilities it needs both internally and with external vendors to support the growth that it desires? Is your PMO looking at static conditions rather than dynamic conditions for its decision-making framework? Does your PMO consider itself the "Center of the Universe" when it comes to dealing with other key groups and vendors?

WHAT WE NEED NOW IS …

What the world needs …

No. Start that over again. What **WE** need is a simple fact-based framework for getting the best truth we can about the world we live in.

For many years, Bob Woodward and Carl Bernstein have been engaged in investigative journalism. Their faces are very familiar because they appear on many newscasts such as CNN and NBC. They are sought-after repeatedly for their very thorough analysis of evolving news including interviews with Washington, DC, figures who make the news every day.

When asked about their methodology, they always remark that they are working to get the "best obtainable version of the truth." That is what investigative journalism is all about. It is fact-based and not opinion-based or emotion-based. That is exactly what the new framework attempts to do in my book *Project Management Lessons Learned: A Continuous Process Improvement Framework*.

This framework is applicable to all scenarios, not just those associated with projects. This framework uses facts, deliverables, and perspectives to give a good overview of what really happened in a specific scenario.

- Deliverables, in a broad scenario, are the tangible things that result from the scenario.
- Facts are those pieces of information that are not refutable in the given situation.
- Perspectives are different viewpoints that participants in the scenario perceive. Often these perspectives must be reconciled by discussion to get the best obtainable "version of the truth."

This framework very much mirrors the investigative reporting that Woodward and Bernstein are famous for.

The framework also makes it very easy to incorporate risk management into the analysis. Since the framework is very "performance-based," it uses the After Action Review questions in its procedure:

- What was the expected result?
- What was the actual result?
- What is the gap?
- What is the lesson learned?

My framework has been employed successfully by some major organizations, such as the Panama Canal, to uncover details of past projects and business situations. This is a powerful framework and often requires the use of facilitation techniques or "reframing" of the scenario details in order to provide results that can be shared with others. Both facilitation and reframing are topics covered by my book, and the reader can research these tools for their own use.

WHAT DOES "ACTIONABILITY" REALLY MEAN IN PROJECT MANAGEMENT?

In my book, I also developed a process for project managers at a Project Close phase of their projects to capture, document, and share lessons learned from their projects with the assistance of the project organization.

In the past, many project groups have attempted to document lessons learned, but inevitably they end up on some shelf in the archives of the group, and never really have any impact on the growth and development of the group's project or business processes going forward. However, I believe that this situation is changing as more and more organizations are embracing the concepts of continuous process improvement. The framework in my book emphasizes that it isn't enough to analyze results or outcomes of the project process and to document lessons learned. The issue of "actionability" is key here.

So, what do we really mean when we use the term "actionability?" It means a number of things.

First, it means that the organization has an intent to capture, document, and share lessons learned.

Second, it means that the organization builds in the capability to capture the lessons learned through a facilitated process in which the project manager, his team, and the stakeholders participate "actively" in the process. There is some debate in organizations today whether the project manager is the best-equipped person to do this. Some organizations are dedicating this activity to a group that is unbiased in its approach, which has the facilitation, framing, reframing, and collaboration skills and competencies to make it work.

Third, it means that the organization builds in the capability to actually implement the changes agreed to in lessons learned statements. This is not easy in some organizations. But what I am advocating is that the lessons learned be written in sufficient detail so that a knowledgeable person in that group who understands the processes and the complexities of the organizational "structure" can implement the changes. Ultimately, this is what "actionability" is all about.

Fourth, it means that the organization devotes tools and mechanisms to sharing the new changes and incorporating these changes in a way that makes them "the business of the company" and not just a "good thing to consider."

"Actionability" is about intent, capture, documentation, and sharing of Project Lessons Learned in a way that becomes part of the fabric of the company's business. If this is not the case in your organization, ask yourself why not. Start today to assess where you are with Project Lessons Learned and what path you can take to make all Project Lessons Learned "actionable" in your business context.

THE "ACCIDENTAL" PROJECT MANAGER DILEMMA

In the field of project management, there are instances of what I term the "accidental" project manager.

This expression can have several connotations, and I would like to give two examples to illustrate the characteristics of the situations.

Once again, perspective is important in determining which connotation you tend to conjure up; depending on your personal beliefs about what a competent project manager should look like; that is, your background, behavior, and performance, you may have a very different idea of what it means to be a project manager.

For example, I have worked with many project organizations that exhibit the full spectrum of beliefs from "Anyone can be a project manager" to "Only those officially certified by the Project Management Institute (PMI) can be considered project managers."

Believe me, there are competent project managers who exist throughout this spectrum.

Typically, we define a "project" as having the following characteristics:

1. Definite start and finish dates;
2. Defined resources; and
3. Schedule, budget, scope, and quality objectives.

Example No. 1:

Recently, our neighborhood's Homeowners' Association purchased a used play and gym set for our neighborhood park area in the subdivision. When they decided to have it delivered and installed, they were faced with the situation that they had purchased it "as is," with no guarantees with regard to moving the set or installing it. So, they decided to look for a project manager among the ranks of the neighbors; that is, someone who could oversee these tasks.

About six months prior to this, a neighbor of mine had orchestrated a backyard wedding for his daughter complete with lighting, music, and the other appropriate support equipment. The wedding was considered very successful in the neighborhood, and the father of the bride was given many kudos for a job well done. As the Homeowners' Association prepared to move the play set to the park, they were informed by several residents about the success of the backyard wedding and the "expertise" of the father in planning and executing the wedding.

The Homeowners' Association approached the father and gained his agreement to be the project manager for the "play" set move and installation.

This is one example of an "accidental" project manager. Typically, it is a person who appears to have the skills necessary to take on an important task, but who has not been truly evaluated for his fit with the new task.

Example No. 2:

Over the past twenty years, as many organizations began to shift from operational and initiative work processes to using projects as a means of converting strategy into action in modern organizations, many of the operations managers were tapped to become project managers who to lead the organization into the new era of project planning and execution. Never mind that many of these operational managers had never heard of project management, a project schedule, or dedicated resources for projects; they were appointed to "make it work."

Most of us who have worked actively in the project management field for many years and have many examples of the second scenario, even in highly respected organizations. The old "chicken" versus "egg" scenario has been implemented many times. Project managers have been assigned responsibilities and roles without having any project management process, standards, or procedures to use. And their training has been sorely lacking.

Typically, what happens in a situation such as this is that management asks the question "Does anyone have any standards for project management that we can use?" And the answer typically comes back "There is a PMI with a website and Certification for project managers."

So, how should project managers be matched with upcoming projects?

One method that I have seen successfully implemented was offered by American Express as a Best Practice through its affiliation with the Corporate Executive Council's PMO Roundtable. They used a spreadsheet with two major parameters: (a) Competency Level of Project Managers versus (b) Complexity of Projects.

The project group would rank the skill and competency level of its PMs at one of five levels. Each project, in turn, would be evaluated versus about twenty complexity factors. Many of the complexity factors would reflect, for example, the number of marketing regions affected, the number of internal company business units affected, environmental considerations, and so on.

Now, this spreadsheet was just a tool that aided Project Group Managers to compare project managers with upcoming projects. The intent was obviously to reduce the risk of making a bad project manager decision. But, that is the case in many of these analyses.

If the Homeowner's Association Board had such a tool, for example, it still might not help the association, however, since their pool of available project managers was lacking. Their problem was just finding available candidates for volunteering.

It is hard to predict the success rate for each project manager in the two scenarios presented. It depends on a number of factors.

I would like to believe that the maturity an organization exhibits in its work processes would be indicative of the way it chooses competent project managers.

How does your organization handle this?

P.S. Now, those of you who have seen the movie "Hoosiers" would probably not consider Coach Norman Dale (Gene Hackman) to be a project manager. No one qualified him as a project manager. But he had a definite start and finish dates with the basketball season at Hickory High School; he had limited resources in terms of a team and coaches (although Dennis Hopper turned out to be a great resource); and he had a schedule, budget, scope, and quality objectives to meet. Could he have been an "accidental" project manager?

HOW CAN PROJECT MANAGERS USE THE SCIENTIFIC METHOD TO FINESSE PROJECTS?

Many years ago, when my daughter was in grade school, the annual science fair was a big happening in the school learning curricula. Students were asked to pick something of interest from science to develop a hypothesis to be proved or disproved and to conduct an experiment that would either verify or negate a hypothesis of their choosing. It was always a big challenge to pick a science fair project that was doable with materials at hand and to define hypotheses that were easily verified or refuted. These science fairs became a family project in the spring each year.

You can certainly recall how these events unfolded. A student would choose to study, for example, tomato plants by exposing the plants to varying amounts of light and then recording tomato plant growth over time versus exposure to light. Varying the light source or making direct lighting versus indirect lighting could be sensitivities to the experiment.

Gradually as our family participated in the annual science fairs, I began to reflect on my own scientific training in high school and college. It brought me back to a revelation that I don't think was part of any science curriculum I ever studied. But it was significant enough to my understanding of science that I kept it in the back of my mind and continued to think about it and its applicability until the present day.

This revelation was what I called a complete scientific investigation or the "scientific method" and I characterized it as a triangle with three topics at the three points. These topics were Theory, Experiment, and Modeling. My "complete scientific investigation" scenario said that "every scientific investigation in order to be complete had to contain elements of Theory, Experiment, and Modeling, and each element contributed to the understanding of the other two."

At the first stage of a scientific investigation, a hypothesis is developed. The hypothesis provides a statement of what outcomes are expected when certain conditions of an experiment are carried out. Experiment is often used to verify or refute a "hypothesis."

Theory is based on observation of the behavior of molecules and compounds under different conditions and is always being updated by new experimental data and often modeling results.

The most famous and publicized example of modeling was that of Watson and Crick in 1953 when they used small "balls" and "sticks" to model the helical structure of DNA. They used experimental data from X-ray crystallography experiments along with theories of how molecules combined when forming larger compounds to determine the shape of the DNA molecule and the spacing/orientation of specific compounds. Several scientists at the time noted that the most important contribution they made to science with that model was the ability to relate several seemingly unrelated theories into a coherent and consistent story of how DNA is formed.

If you are a fan of one of the CSI (Crime Scene Investigation) TV programs, you know that evidence from fingerprints and DNA has been used extensively to put together profiles and complete descriptions of the subjects.

As a project manager or a PMO team leader, what might you gain from this simple understanding of the scientific method?

First, developing a hypothesis about cause and effect, behavior/activity, and performance/outcomes would be a good starting point to analyze project and PMO activity.

Second, theories about the interaction and collaboration of teams and new process development might be a good starting point to understand what might unfold in a project.

Third, developing the ability to put together several seemingly unrelated pieces of information might be a good development target. For example, listen intently to what your sponsors and stakeholders are saying about your deliverables and your project progress. It might be just the specifics that help you meet their objectives.

Fourth, experiment whenever it makes sense. Put two team members together who have not worked well in the past. Pull in some research or benchmarking with other groups to infuse your project thinking.

Fifth, tie modeling to "experiment," "experiment to theory," and "modeling to theory" in a package that provides as complete a look as possible to your project or PMO initiative.

You will be glad you took the time to see how scientific investigation can lead to better projects and outcomes.

WHEN IS PROJECT MANAGEMENT MUCH MORE THAN JUST COST, SCHEDULE, SCOPE, AND QUALITY?

Some project managers reading this blog are engaged in developing software, hardware, and systems applications for corporate or other uses by stakeholders who bank on their technical expertise to provide solutions for cost, scope, quality, and schedule issues.

Other project managers are engaged in designing and building "brick-and-mortar" facilities.

Still others create value for organizations by providing project deliverables that satisfy requirements for new transactional systems applications that enable the business processes of organizations to realize their full effectiveness.

In general, the project business requirements for the project managers cited above dictate an in depth understanding of stakeholder behavior under the most rational of conditions and those stakeholders usually recognize the importance of a collaborative effort with the project manager in defining project business requirements.

However, there is another set of projects for which the requirements space is not so rational, the stakeholders are often distraught and emotional, and the path forward to a completed project deliverable can seem far away. In times of a fire, flood, emergency, or disaster, the person who suffers the loss of property and close personal assets can often be devastated and not think in a rational manner. Who do I turn to? How do I recover? What is my best alternative? These questions are foremost in the minds of the people immediately affected by a disaster loss.

And the project managers who renew assets devastated by fire, water, tornado, or other disasters often deal more with the human, emotional, and relational elements in their roles as project managers than they do in the normal transactional elements of projects.

I visited with John Esposito and his team at ESPO Fire and Water in Tulsa, Oklahoma. ESPO Fire and Water is engaged in restoring and renovating properties damaged by fire, water, and other secondary entities such as mold. John's team is often first on the scene of a disaster because they are often referred by others who have experienced the same calamitous conditions.

I especially wanted to study their project management process because of their unique requirements to deal firsthand with the relationship aspects of customer service in the most demanding of conditions. This is a view most of the project community does not normally see in their everyday project environment.

Kevin Hefley and Ann Parker are ESPO project managers with over twenty years of experience each in fire and water damage assessment and clean up work. They are extremely well versed in the jargon of scheduling, estimating, planning, budgeting, and quality which the project management discipline brings to a business like ESPO. But, they keenly understand that the relationship aspect of customer interaction is far more important in their business context than the transactional aspects.

Kevin Hefley told me, "Often the first person on site in a disaster, fire, or flood situation spends most of his time with the distraught victims who are trying to grapple with the situation. We know that other ESPO responders must play more than one role in these cases so we can quickly calm the victims and assist in their thought processes." You certainly don't find this type of project management training in the PMBOK or other traditional project training.

"And often it is a case," he went on to say "of making the customer aware of standards that must be met in the remediation. There are ANSI standards (American National Standards Institute) like mold standards."

Kevin said "Every morning before I begin work, I refer to a Chinese proverb on my computer which says 'Before you ask yourself if you are doing things right, first ask yourself if you are doing the right things.' It has become a philosophy with me because we are often acting so fast in dealing with multiple projects."

As I talked with Ann and Kevin, I realized that the process they were following was extremely mature, the result of more than twenty years of understanding what works and doesn't work in disaster situations. Of more than twenty years of learning from others in the organization and those people in turn learning from them.

Ann commented on the importance of the customer at every step. "There was one job where the customer wanted to know exactly what we planned to accomplish each day on the job and, since that customer was on my way to work, I stopped by each morning to make sure the customer knew exactly what would happen that day."

Kevin and Ann rely heavily on vendors to supply basic work like electrical and plumbing. Management of these vendors is facilitated by the fact that most of the vendors have worked with them for a number of years so they understand the quality work being done.

The technical director of ESPO Fire and Water is actually a member of the ANSI standards board so he has a very interactive role to play in both setting new standards and interpreting existing standards.

I also asked Ann and Kevin whether they used best practices in their project management. They described to me an extremely mature business process which relied heavily on best practices—even including preparing the trucks each night so that they would be ready to roll at a moment's notice (which is all you get in a disaster situation)! Similarly, the technical director of ESPO Fire and Water is a member of the ANSI standards board. He has a very interactive role to play in both setting new standards and in interpreting existing standards.

In Michael Hammer's classic book *Beyond Reengineering*, he discusses the fact that the natural consequence of an organization focusing on well-defined processes and becoming process-centered is "professionalizing" the work. In this context, Michael Hammer says:

> Three words characterize the worldview of a professional: customer, result and process. The professional sees himself or herself as responsible to the customer; the mission is to solve a problem for the customer, to create the value that the customer requires. If that value is not created, if that problem is not solved, the professional has not done his or her job. It is only by producing the result that the customer requires, by performing the entire process that yields that result, that the professional discharges his or her responsibility.

Other scholars have added to this definition by insisting that the professional is one who is continuously learning on the job and reflects on his experiences while sharing lessons learned with others in his practice.

Based on my experience working with a number of project management groups in large and small organizations, I would have to say that there is a high degree of professionalism displayed by the project teams at ESPO Fire and Water. I would say that any other such organization wishing to evaluate their own internal processes could use ESPO as a benchmark in that respect.

As Ann Parker says "The greatest satisfaction I get from my project work is when a customer calls me six months after a completed job and says literally 'thanks for giving me my life back.'"

I had to share this with the project community because it proclaims an element of caring in project management that we often do not see in our daily project work.

What level of professionalism does your project organization exhibit?

Is your process driven by transaction or relationship?

THE LITTLE SANDALS THAT COULD

In February 2012, my wife, Linda, had a knee replacement surgery at the Mayo Clinic in Phoenix, AZ. We planned to be in Phoenix for at least six weeks following the surgery for physical therapy and follow-up visits at the Mayo Clinic. As we packed our bags for the trip to Phoenix, Linda included a new pair of white New Balance shoes for her physical therapy. Toward the end of the packing, she also packed an old, ragged, pair of brown sandals with the idea that the weather might be warm enough to wear some really comfortable shoes around our hotel.

Linda's surgery went well with no real complications. But because she had experienced a DVT (Deep Vein Thrombosis) in her left leg about thirty years ago, the doctors were very cautious about her care and even installed a valve in her vein system to keep any clots which developed in the legs from traveling to the heart. Her left leg, ankle, and foot were extremely swollen from the surgery and remained that way for about six months.

When it came time for physical therapy, Linda realized that she could not get the new, New Balance trainers on her swollen left foot. What to do? As we thought about alternatives, the little pair of brown sandals seemed to be peeking out from a corner of the luggage. She wore those sandals to physical therapy that first day, and for every day after for about five months. They seemed to be the most appropriate alternative for her swollen foot and for maintaining mobility.

When I stopped to think about it, I was amazed that we had almost excluded the little brown sandals from her Phoenix wardrobe. But I also thought that maybe we hadn't taken into account some risks from the surgery, such as a swollen leg, ankle, and foot.

So I dubbed the brown sandals "The Little Sandals That Could."

When we returned to Tulsa, Linda purchased two other new pairs of sandals to wear. But she continued to rely on the little sandals that could.

Now what in the world does this have to do with project management, you may ask?

Ask yourself these questions:

How many project assets do you have in your portfolio that you may not be using to full advantage in your project work?

How many members of your project team or your local project community have you overlooked, who might have skills and capabilities that might lead to real success on your projects?

Remember "The Little Sandals That Could." Are there "risks" in your project work for which some lesser project assets might provide good mitigation tools and accessories for alleviating and dealing with project risk?

William Bridges once said, "Change Creates Opportunity." Change creates interfaces between old values and new values, between old methods and new methods, and between old processes and new processes. Those people who can successfully bridge the gap between old and new either by communication, facilitation, interpretation, or translation can play a key role in "Change Success."

How many "Little Sandals" are there in your organization who can fit this bill?

VISUALIZING YOUR FUTURE IN PROJECT MANAGEMENT

Consider the meteoric rise of Dunkin' Donuts in the last few years. Its brand is everywhere to be found now.

I recall a time when my daughter Christina was a student at Yale University in New Haven, CT. My wife and I would often fly into JFK in New York City and drive the I-95 to New Haven. Along the way, we always made Darien, CT, a favorite stop. Darien was a sleepy but affluent community along the I-95 and the AMTRAK tracks to New York City. Each time we would plan to stop at the Dunkin' Donuts shop on the main street in Darien. The shop was right next to the AMTRAK station so customers were always coming and going on their way to or from New York City. We would always think about what type of donuts we would have, and I could just taste the great coffee when I was driving I-95. Would it be a jelly-filled donut and what flavor? We would always take a little bag of goodies to Christina who always claimed hardship with the food in the Morse College dining hall at Yale.

It was fun to watch the patrons of the donut shop rush in to place their orders and then run to catch a train. Or, in the evening when we were on our way back to JFK, we often stopped at Dunkin' Donuts for a quick donut and coffee. People always seemed to be de-training from their trip from Grand Central Station to be picked up by a family member but also to stop and get a donut and coffee.

Someone in that organization had to visualize what it would be like if they could someday rival Starbucks in name recognition. If somehow, the customers that frequented shops like the one in Darien would take that name Dunkin' Donuts to a new height in the world of business. Lou Tice teaches us a basic principle that "People move toward and become like that which they think about."

Someone thought about what Dunkin' Donuts could be. They visualized how it would work. They gathered the impetus from their previous successful experiences and made Dunkin' Donuts the name it is today.

Today the brand "Dunkin' Donuts" is everywhere. Their coffee is sold not only in their stores but in other retail stores on the same shelves as Starbucks. Their brand presence is the result of someone clearly visualizing what a future might look like for such a brand and then proceeding to plan for it. Goal setting, expectations, performance measurement, and feedback have clearly become elements of their continuous performance improvement framework. They are a living, breathing BRAND.

Lou Tice, who I have mentioned previously in this book, was a great teacher, mentor, and motivator whose principles have been passed on to individuals and organizations around the world and have created successful lives wherever they have been applied. I really believe in his concepts and believe they apply to a vast array of human experiences in business, social, family, and societal contexts.

He introduced me to a method of visualization and affirmation which I have used for many years. It is all based on the principle that you move toward and become like that which you think about and that your thoughts and words control your future. It is also based on the principle that people act in accordance with the truth as they perceive it to be.

Visualization is exactly the method that got you where you are today. What you have achieved, what you currently believe, and what aspirations you currently have were all conceived as you thought about and planned a path for your future. It is true that many people take paths in their lifetime that change directions or that were influenced by factors that they reacted to. But ultimately you are where you are today because you visualized what you wanted.

Lou Tice often pointed out that the amateur leaves to chance what the professional controls in this arena. In other words, the professional uses clear visualization and affirmations to stir their subconscious to create a future that they envision. Dunkin' Donuts had a workforce using clear visualization to get them to the brand powerhouse that they are today.

One of the distinguishing features of Lou Tice's treatment of visualization and affirmation was his explanation of the "states" of being. For every aspect of life for which an individual sets a goal or objective, such as business, social, spiritual, intellectual, professional, and so on, there are any number of states of likely possibility for that aspect of life. For example, suppose a person sees that his current situation is such that he cannot provide some of the finer things in life like vacations and special gifts to his family but his desire is to financially be able to do so, he can set objectives and goals for achieving that situation or "state" and then write affirmations and visualize so that the subconscious will create that new "state." There are of course many possible "states" along the way between the current situation or "state" and the end objective "state."

But often it's the case that a person will look at his checkbook or bank account and realize that this current "state" is far from adequate to do the things he desires to do. The mistake is to focus on the current state because the subconscious will create that "state" continually by recognizing that it is the one being focused. The key is to affirm and visualize movement to a higher level "state" in which the objectives and goals are reached. If this is the case, the person will create a subconscious energy to find the resources and the knowledge to reach that end "state."

What does this have to do with project management? If you are a project manager who continually sees that your progress toward your professional or personal goals is not being met, you can adopt the visualization and affirmation methods of Lou Tice. I highly recommend that you spend a good amount of time assessing your goals and objectives in every major aspect of your life before you proceed with this course, however.

Controlling your talk is very important in the affirmation and visualization process. Self-talk is that three-dimensional form of words, thoughts, and actions that goes along with your daily activity. In the normal course of interacting with others, we talk and then tell ourselves a story about how are feeling about the activity we are engaged in. Or we react to the input of others and we tell ourselves a story about the activity. Self-talk is very important in the development of the right self-esteem that really controls our achievement of goals and objectives.

> Our present thoughts determine our future. So, if you want a better future, begin today. Use the power of your mind to first see the change, then live it. There is great power in purpose.
>
> Louis Tice

THE ROLE OF COGNITIVE SCIENCES IN PROJECT MANAGEMENT AND PMO PERFORMANCE

In the past seventy-five years since the Manhattan Project in World War II, there have been tremendous advances in the way project management has been conducted resulting in increases in performance, deliverability, repeatability, and consistency in delivering results. This increase in performance has been the result of the introduction of computing, information, and intelligent systems, the development and use of information and knowledge systems for decision-making, and the evolution of cognitive science processes for human performance.

In addition, the introduction of the Program or Portfolio Management Office (PMO) as a Best Practice in larger, more complex organizations has enhanced the organization's ability to juggle huge demands in project delivery and organizational success. Demands on PMOs have rendered them essential to overall business success and growth of business volume.

Cognitive science has emerged in the past fifty years as a key component in the success of project managers to manage complex, multidimensional projects successfully.

The Latin root "Cogn-" means "to learn." So cognitive science deals with processes that relate to learning, knowledge, and how knowledge is assembled and utilized. This include the following:

1. Attention (and osmosis)
2. Creativity (creative subconscious)
3. Language and linguistics (use of "milestones," "stakeholders," etc.)
4. Problem solving (and risk management, innovation, and design thinking)
5. Reasoning (and resilience)
6. Perception (lessons learned)
7. Memory (knowledge management)
8. Intelligence (numerous topics)
9. Reflection (lessons learned)
10. Communication (and best practices, benchmarking)

The simple fact is that the cognitive sciences increased the overall capacity of project managers and the PMO to meet the challenges of modern business and its complexities.

In summary, the cognitive sciences field developed as a break from behaviorism in psychological thought. It has had a tremendous impact on the advancement of project management and its contribution to modern society.

TOWARD A PROCESS STATE OF MIND

I am sure that most of my readership is familiar with the Billy Joel song "New York State of Mind." Those of you around New York will recognize in this song some familiar landmarks and favorite places which he says create a "New York State of Mind." His longing for the old familiar of New York and its surroundings is particularly appealing in this song, which has become one of my favorites.

I would like to use Billy Joel's theme to introduce my readers and friends to a concept I call "a Process State of Mind."

As I talk to friends and colleagues in the corporate world or who own their own small consulting firm or who are engaged in activities in their communities every day, one thing they are all craving is discipline and structure and organization in their everyday lives and work.

They are all familiar with the concepts I have been writing about for the past few years in terms of PROCESS, PROJECT, and LESSONS LEARNED. And they have asked me for more description of these practices in many cases.

While most of my writing and my book on Project Lessons Learned has focused on lessons learned and feedback to improve project management processes, these same concepts can be applied to any PROCESS. PROCESS is, I believe, the very discipline that my readers and colleagues are seeking and need in their work and their lives.

A PROCESS is a set of activities or tasks which, when performed in a prescribed sequence, yield a result or an outcome and which can be improved.

Yet, a process without some mechanism or format for improvement is not really a PROCESS.

As I talk to friends and colleagues, some basic questions come to mind that will make it easier for them to incorporate PROCESS into their everyday activities, so I would like to address some of these issues to bring us to a real "process state of mind."

First, who should be involved in making process improvements to a process? Ideally, the participants who plan and execute the process are best equipped to identify, document, and share "lessons learned" from the process, which can become "actionable" process improvements.

Second, if process improvements identified by the process participants seem to be few and far between, it may be that the process is coming close to optimization, although this state is hardly ever achieved due to the "dynamic" environment that most processes exist in. Owners of the process should seek further improvements through Benchmarking or Best Practices.

The principles I have applied in my book on Project Lessons Learned can be applied to continuous process improvement scenarios.

But don't just take my word for it—here's another example of process improvement being employed in a discipline totally different from project management:

Doris Kearns Goodwin is considered by many to be "America's historian." She has written authoritative books about the Kennedys, Roosevelts, and Abraham Lincoln. But she will quickly tell you that the most asked questions in her lecture tours are about her childhood and the Brooklyn Dodgers.

She was born in Rockville Centre on Long Island, New York. Her father was a financial services employee who worked in Manhattan during the day. His desire to learn about the daily games of the Brooklyn Dodgers led him to teach Doris about the game of baseball by using a small red scorebook to keep score. He taught her the shorthand symbols for scoring the game, such as "K" for strikeout and "1-3" in the scoring section for an inning (to indicate a put out by an infielder after a ball hit by the batter to the infield).

Each afternoon Doris would listen to the radio broadcast of the Dodger games and record in her scorebook the events of the game. In the evening after dinner, she would sit with her father and recount the baseball game of the day.

"From something as simple as the small red scorebook in which I inscribed the narrative of a ball game, I saw the inception of what has become my life work as a historian," she says in her book *Wait Till Next Year*.

> Through my knowledge, I commanded my father's undivided attention, the sign of his love. It would instill in me an early awareness of the power of narrative, which would introduce a lifetime of storytelling, fueled by the naive confidence that others would find me as entertaining as my father did.

Doris employed a PROCESS to achieve her objective of bringing the daily baseball game summary to her father. She did not call it PROCESS then, but that is exactly what it was, a disciplined and organized way of achieving an objective each day.

And the little red scorebook we would term a PROCESS TOOL in today's process vocabulary. With the scorebook, she could answer her father's questions such as "How many strikeouts did Don Newcombe get today?" or "How many hits did Roy Campanella get today?"

In her process, Kearns Goodwin probably employed such questions as these to help her define the specific activities making up the process:

- What is game time?
- Is my radio tuned to the right channel?
- Do I need any other paper and pencil to make notes during the broadcast?
- Are there any terms Dodger announcer Red Barber has used recently that need clarification?
- Will my father and I review the game at the same time and location as usual?

This PROCESS served Doris Kearns Goodwin well in meeting her objectives. And I am sure that she identified process improvements along the way. At one point in her book, she recounts that at particularly tense moments in some games, she would actually mimic the vice and expressions of Red Barber to add realism and interest … a form of process improvement.

So let's create a "Process State of Mind" by looking for every opportunity in our organizations and in our lives to define processes that lead to meaningful outcomes and which can be improved. The structure and organization that this will introduce to your daily activities will give you a good feeling for day-to-day living.

WHAT A REVELATION!!! BEHAVIOR DETERMINES PERFORMANCE

For over twenty years, I have worked on Project Lessons Learned in organizations. There are two different forms of Lessons Learned that project managers and PMO practitioners must understand to fully comprehend the concept of lessons learned.

The surface connotation of lessons learned deals with projects as they turn "strategy" into "action" in an organization. As projects create results and outcomes, sometimes these outcomes are not what was desired by the project. Or, there may be deliverables that don't meet expectations. These projects are good candidates for process improvement in the form of lessons learned. My book *Project Management Lessons Learned: A Continuous Process Improvement Framework* provides a great framework for identifying, capturing, and incorporating lessons learned in the project process. Over time, I have also learned that this framework is extremely useful as a fact-based investigative framework that can be applied to any performance-based scenario. This becomes highly important in today's world where we are faced with a virus pandemic and a myriad of situations needing examination.

But the subtle thing about performance is that it takes "behavior" to reveal the end product or the performance we desire. *Doesn't it make sense that we should examine every aspect of "behavior" in order to realize the best performance we can achieve?*

These are the surface, day-to-day, most identifiable lessons learned. Although corporations have been talking about lessons learned for many years, much of what was learned about lessons learned in the early years was forgotten because the lessons learned were developed in a culture of "Blame." If an organization thought it was useful to capture lessons learned, it was in an environment where finger pointing and attaching blame for what went wrong with the project. Over the years as companies, organizations, and corporations became more enlightened about what process improvement could do to move the organization forward, surface lessons learned became more accepted in organizations and much effort was devoted to rigorously identifying and making accessible lessons learned to everyone who could benefit from them.

As I've discussed previously in this book, I traveled to Panama in 2011 at the request of the Panama Canal Authority to facilitate a project management lessons learned process in support of their five billion dollar Panama Canal Expansion Project.

The concept of the surface lessons learned has even seemed to mature in recent years and incorporated directly into the project management process. At the end of each phase gate, a lessons learned review now seems to be the standard before the project proceeds to the next phase.

I have also encountered a lessons learned phenomenon embedded in the organizational framework that is related to how management and organizational goals and policies are developed and implemented in an organization. The age-old question "Does structure influence behavior?" is addressed by this lessons learned framework. I first recognized this "structure influencing behavior" framework while working for Ford Motor Company many years ago. Today, the area of organizational dynamics and systems thinking is unlocking many observations and lessons about how organizations introduce behaviors through their management goals and processes, which can be identified and changed to improve organizational performance. Daniel Kim in his book *Introduction to Systems Thinking* covers this framework by introducing an overlying framework involving "Vision," "Mental models," "Systemic Structure," "Patterns of Behavior," and "Events."

Despite what we as project managers have learned about PMBOK and specific AGILE methodologies, I believe "behavior" determines performance in a project team environment. Examples include project teams sharing information in the project environment with other project teams, assigning diverse project team members to teams to foster innovation, sharing stories between project team members and project managers as a learning technique, identifying best practices and engaging in benchmarking processes, fostering clear communications among team members and the project manager, looking for ways to lessen stress in teams during crises, and so on.

Performance, performance, performance! That is what every PMO is clamoring for. But it is "behavior" behind the scenes on the part of project managers, team members, stakeholders, sponsors, and vendors that really determines the performance and success of projects.

So, as you continue your career completing projects and advancing the methods of the project management discipline, always keep in mind that "awareness" of behavior is so important to the overall success of the discipline.

BUILDING FUTURE PMO TECHNOLOGY DEVELOPMENT "CAPABILITY" IS THE KEY TO ADDING "VALUE"

As I work with project managers in many fields, it is becoming more apparent every day that, in the future, new technology development will play a greater role in delivering "value" from project management. Of course, new technology development has been a major part of many projects. Witness the development of electrical components and control systems in the Panama Canal project from the early 20th century, as I discussed in my *Lessons Learned* book. Today, we only have to look at the technology and new business fields, and such sources as CNBC TV and CNN, to hear about new "apps" and new technologies that are adding new, or replacing old, functionality in products and services.

Look no further than Walmart's growth into e-commerce and Amazon's growth into more brick-and-mortar solutions to find that new technologies are at the heart of business and consumer demands.

So companies are struggling with the question, "Do I grow this new technology by acquisition or by internal development of the skills, talents, and structure to compete in the future?" The answer to this question lies in another question:

"How do I develop the 'capability' that will enable me to deliver my desired end value?"

This was much the same question many years ago when organizations began to concentrate their project management skills, practices, and methodologies into an internal group that became known as the "Program Management Office (PMO)." The PMO quickly became a standard and "Best Practice" for delivering consistency and repeatability of project results. The PMO is often referred to as a "capability-based" organization because it was constructed to DELIVER projects successfully.

CAPABILITY is the combination of people (resources, talents, leadership); processes; technologies; and organization (structure, networks), which allows an individual or organization to deliver its intended objectives. The blueprint covers all those components, but not separately. It covers how they will fit together. There is also an accompanying plan that specifies the people who will build pieces of the capability, the targets, and incentives that govern their actions, and a timetable for the implementation.

Recent studies by Booz and Company consultants have focused on "capability building" as the key to closing what they refer to as the "strategy-to-execution" gap that has plagued companies for many years. Many times in the past, a great strategy never seemed to materialize into actionable results. This was termed the "strategy-to-execution gap." Their analysis has been documented in the 2016 book *Strategy That Works: How Winning Companies Close the Strategy-to-Execution Gap* by Paul Leinwand and Cesare Mainardi, with Art Kleiner. Their process for building new capabilities focuses on answering the following questions:

1. What is the capability?
2. Why is it valuable?
3. How would it be different from what we have today?
4. Describe a day in the life of this capability; what does it look like?
5. What is required to make it work?
6. For the "capabilities system," what does the business case look like?
7. How does this capability fit with others in the capability system?

The focus of this analysis was to build meaningful and lasting capabilities and to foster cross-functional relationships between corporate competencies that complement each other. In short, their theme was "build capabilities before seeking results."

To summarize, PMOs in the future will need to develop new technology capabilities systems which will add to their project delivery capabilities. Of course, specialized PMOs in companies such as biopharmaceuticals or information systems will need to consider industry-specific technologies to insure their "table-stakes" status. This new reality may require a "maturity curve" approach in implementation to be successful in the long term.

DRIVING INNOVATION IN PROJECT MANAGEMENT: THE NEW PARADIGM

Those of us who have been actively engaged in project management for many years know that it is an evolving discipline that has expanded its tools and techniques as new challenges have arisen. Today one of the most pressing issues facing practitioners in project management is the need to introduce innovative techniques into all facets of project management.

In recent years, innovation has been a catch-all term for all things related to looking at new ways of conducting project management practices, and turning ideas into value for stakeholders. *The Medici Effect*, a book that was first published in the early 2000s, and which was recently re-released with additional content, addresses this interest in innovation.

The basic themes of *The Medici Effect* are:

- Innovation is more likely to occur at the intersection of two disciplines or fields.
- Diverse teams drive innovation more than homogeneous teams that are composed of team members with very similar backgrounds and experiences.

There are three reasons why *The Medici Effect* has been so pronounced in introducing innovation:

- The Movement of People. As more people have moved from region to region and country to country, cultures have mixed in terms of ideas and beliefs/values being shared and impacting behaviors. The perspectives of people have changed as they interact and understand how they can cooperate.

- The Convergence of Sciences. As scientific fields have been divided into more specialties, the intersection of these specialty fields has given rise to more combinations of solutions for problems facing society.
- The leap in computation capability. As more data is being analyzed with faster and more complex systems, the relationships among scenarios, decisions, actions, and behavior of people are being studied and analyzed. The result is a surprisingly high accuracy of behavioral understanding and a lessening of risks associated with predictive techniques.

A good example of "innovation" arising at the intersection of fields is the scenario at Dunkirk in 1940, the early days of World War II. The British Expeditionary Force was trapped on the beaches of Dunkirk in France just across the English channel from England. The Germans were advancing toward the beaches from inland and would have overtaken and captured the British Forces. There was a limited number of British naval vessels in the channel at the time to move the British Forces from Dunkirk to the English mainland.

This problem demanded some extraordinary innovation and thought processes to affect a solution. The British military called out to all British citizens who owned pleasure boats moored on the English side of the channel to sail to Dunkirk with haste to help rescue the British Expeditionary Force.

This thinking represented an intersection between the military and private citizens of Britain. The rescue was successful.

In my book *Project Management Lessons Learned: A Continuous Process Improvement Framework*, the concept of an "innovation PMO" is introduced to describe an organizational framework in which the PMO is focused on turning new ideas into reality in terms of products, services, or new processes. Often, when existing PMOs evaluate their capabilities for developing new innovative technologies, the focus evolves into understanding and implementing new "capabilities" in terms of processes, organization, technology, and people. It is the responsibility of management to develop Competency and Capability for these innovative thrusts in their business context.

In the Booz and Company publication *Strategies That Work*, they strongly emphasize that their studies have shown that initiating cross-functional teams which bring into play many different disciplines within an organization is a key to innovative project thinking.

The nature of business today involves disruption of existing business models and plans by new applications of technology which destroys or threatens the existing business models. Take the real estate market, for example. The old process of listing a home, preparing it for sale, entertaining potential buyers, and handling the sales process has largely been replaced by new technology models such as OfferPad.com and other processes that emphasize a reduction in home handling processes by real estate agents. Many online order businesses have disrupted traditional brick-and-mortar businesses such as Walmart, Target, and Home Depot.

I am encouraging all project managers to look for these intersections of fields and interests to unlock innovative potential in their project teams. They should seek out diversity when organizing new project teams because it brings unfamiliar viewpoints and perspectives to the problems facing teams. Actively seek out new intersections of fields and disciplines in your project work.

WHAT CAN PROJECT MANAGERS LEARN FROM THE MOVIES?

Before Valentine's Day a few years back, I was browsing through the movie/DVD section of my local Barnes & Noble looking for nothing in particular. A display of Valentine's Day suggestions caught my eye. One DVD stuck out because it seemed particularly well-priced, and the theme was one we all welcome on Valentine's Day—a great love story between two seemingly very different people, Elizabeth Bennet and Mr. Darcy.

The DVD was "Pride and Prejudice," starring Keira Knightley, Matthew MacFadyen, Donald Sutherland, and Dame Judi Dench. This 2005 film was nominated for four Academy Awards and was called by many "the Best Film of the Year." It is, of course, a love story based on the Jane Austen book *Pride and Prejudice.*

This particular DVD contained not only the movie but several additional features: A Bennet Family Portrait; Jane Austen: Ahead of Her Time; Behind the Scenes at the Ball; "Pride and Prejudice" HBO First Look; and Commentary from Director Joe Wright.

As I watched the movie and several of the other features over Valentine's Day weekend with my wife, it struck me that the entire package created a tremendous value when viewed as a whole. As I began to think about the plot of the movie and the way it unfolded, I could not help but think of some similarities with the way projects are planned and executed.

In our experience as project managers, it is too often the case that the project manager wants to finish a project as quickly as possible and then get on with the next project. After all, that's what project managers do, manage (and complete!) projects. Too often in the past, the project manager was focused on the sponsor as the principal stakeholder in the project, and often—unless a very robust OCM plan was included—the project manager did not focus so much on the end users as "real" stakeholders.

This DVD package, however, considered all the stakeholders (i.e., the various possible viewers of the DVD). The movie, when combined with the DVD "extras," answered questions about how the love story unfolded, gave helpful background information, and additional information that showed viewers multiple ways that they could understand or "frame" the story. In doing so, the DVD anticipated and answered questions that the audience (the buyers of the DVD) probably had not considered until after having purchased and viewed the movie itself. How many projects do the same thing? Probably not many these days.

So, as you proceed with your next project, consider some of the following lessons in capturing all the value and satisfying all the stakeholders. Pretend that you are the "Director" and charged with telling the story as well as creating all the "behind the scenes" explanations that stakeholders want both explicitly and implicitly:

1. "Frame" the plot and the interactions of the characters correctly. Look at the social or business context in which the action takes place and make that context real (i.e., come alive).
2. Assemble the right cast, obviously the cast members must complement the entire context of the project.
3. Train the cast in the "context" of the specific project and its tasks. Each cast member brings some "raw talent" to the party but the director's task is to mold that talent in the context of the movie so that the value flows from the performances of each cast member as they interact, as they see and feel the action.
4. Consider all the stakeholders, including the end users, and focus on the "needs" of the end users; not just the "expressed" needs but the underlying needs.
5. Consider not only the sponsor's vision for the project but also the process by which the project will be executed. After all, a director's vision for a movie must not only follow the screenwriter's narrative but also the

process for the movie production. This is also where the PMO comes into play. Having a defined process for all projects means all projects are initiated, justified, authorized, executed, and closed by a process which can be repeatable and consistent. That said, it does not mean that every project must conform exactly to the process, especially if there are situations where the process doesn't exactly apply. It means that there are "guidelines" for how to conduct the project (or to develop the movie) which have contributed to success over time. The PMO is a guiding standard for reference and for consultation.

6. Identify and address any risks to carrying out the plan for the project as originally conceived and have adequate plans in place to address those risks.

The next time you view a motion picture, ask yourself if the director fully accounted for all the stakeholders' wishes in its production. And the next time you plan a project, think of the lessons we can learn from another art form such as a motion picture.

For example, I learned a lesson from Director Joe Wright's statement in his narrative of the film that "the experience of making this film informs the film itself."

In this context, the verb "informs" means "to give evidence, substance, character or distinction to."

How many of us as project managers would ever say that "the experience of managing this project informs the project itself?" And yet, if you really think about it, every project manager adds substance, evidence, character, and distinction to his or her project by managing the project every day … from the inception of first thought to the final close.

That is the key to successful projects. To the extent that the PMO can pass along this type of ownership of the entire project experience to project managers whose experience will inform the projects, we will see more and more successful projects.

WHAT IS LEVERAGE AND HOW CAN PROJECT MANAGERS USE IT TO THEIR ADVANTAGE?

Greek mathematician and inventor Archimedes said, "Give me a place to stand, and with a lever I will move the whole world."

As a project manager, you have probably used the term "lever" in the engineering or physics context to describe a device that can be used to gain a "mechanical advantage" in a given scenario. It means that you can apply force, in varying amounts, to perform useful work.

Likewise, "leverage," a commonly used word derived from the word "lever," means the application of any number of concepts to achieve some advantage in a given scenario for the person or group that is skilled in the application of the concept.

Some forms of leverage might be an individual or group's position in an organization, a specialized knowledge, a unique piece of information, robust technology, and so on.

A good example of technology as a leveraging variable arose in a New York Yankees baseball game with the Detroit Tigers. With Yankee Mark Teixeira on third base, the batter hit a ground ball to the infield that resulted in a throw from first base to home to stop a run from scoring. Teixeira slid into home plate and touched the base with his left hand while the Detroit catcher was almost simultaneously taking the throw from first base and applying the tag to Teixeira.

The home plate umpire called Teixeira out, but Yankees Manager Joe Girardi disagreed. He challenged the call while his Yankee bench coach called the press box to get some other camera views of the play. Under the challenge, the home plate umpire and umpiring crew chief consulted with the instant replay staff, whose use of camera technology has become an accepted practice in Major League Baseball.

The challenge by Girardi was upheld, and the home plate umpire's call was reversed.

Girardi had used technology to leverage the situation.

Prior to instant replay technology and challenge, the play would have resulted in an out for the Yankees, and no run scored.

Think of areas in your project management experience where you or a project team member, stakeholder, or third party has used position, knowledge, information, or technology to leverage a situation. How might you have prepared yourself or your team to leverage your outcomes and your project objectives?

In my book *Lessons Learned: Taking Project Management to a New Level in a Continuous Process Improvement Framework*, I have also applied leverage in an organizational dynamics connotation to describe how Project Lessons Learned can be derived from seeing single projects as events but recognizing that a more leveraging connotation for project groups can be identified. This can help you make lasting changes in project or business processes by identifying patterns of behavior among a group of projects subject to the same project environment.

"Leverage," in this case, is insight into how project teams behave, how they organize themselves, what management principles and objectives they choose to follow, and what they value as being the truth for their direction.

If you as a project manager have not considered "leverage" in your day-to-day work process, give this some thought.

You might turn out to be the "lever" that Archimedes referred to who can move the whole world.

WE HAD EVERYTHING TO GAIN ...

"We had everything to gain by planning and working closely together to advance the development and maturity of our new IT Project Office, but it seemed that every time we would take a step together in the right direction, one of us would sideswipe the accomplishment by acting irrationally or in what seemed like an irrational manner."

Those were the words of my former manager in an IT Project Office (a predecessor to the modern PMO). It is an often-repeated statement, but one that does not get much scrutiny from a root cause analysis format.

Here is the scenario:

A major energy company acquired the downstream assets of another energy company and integrated the functional groups, including the Information Technology groups. At the start of this scenario, I was a Project and Planning Consultant in the IT Planning Group. The Information Technology group decided to form an IT Project Office (ITPO), and it hired an experienced manager from a major Fortune 500 Company whose expertise was in forming and maturing PMO-type organizations. Prior to that point, each IT Applications Development, Systems, and Infrastructure Group had planned and executed projects within their own groups with limited collaboration across groups.

A major consulting firm had facilitated the merger of the two energy companies, and management of the merged company strongly recommended that the Information Technology group use the consulting firm as a "guide" in forming the ITPO. The consulting firm had an excellent reputation for internal project management capability, and it utilized a methodology which I will refer to here as "THE METHOD." So, Information Technology Management was pleased at the outset that they not only had an experienced PMO-type manager but also a strong consulting group direction.

The "vision" and directions given by the IT Project Office Management to the consulting firm were that the ITPO wanted to instill its own "trademark" and "business context" in the new project group.

While the consulting firm heard this firm direction, it internally recognized that the firm had made a great investment in "THE METHOD" and that it would exploit "THE METHOD" at every opportunity.

The ITPO utilized the SEI Capability Maturity Model as a framework for planning its path of evolution to a mature state. However, whenever a new process or procedure was developed in concert with the consulting group, the outcome had a strong flavor of "THE METHOD." So, whenever the ITPO mapped its systems projects to follow the business processes with which it was trying to align, it conveniently left the consulting group out of the process until some redefinition of an omitted process was imminent.

This recurring pattern of behavior was subtle but highly visible to those of us living the daily ITPO experience.

This is an example of a "systems archetype" at work.

Peter Senge has written extensively about organizational dynamics and behavior and systems archetypes identifiable from events and patterns of behavior.

William Braun has also written extensively about systems archetypes. System archetypes are highly effective tools for gaining insight into patterns of behavior, themselves reflective of the underlying "structure" of the system being studied. The archetypes can be applied in two ways—diagnostically and prospectively.

Diagnostically, archetypes help managers recognize patterns of behavior that are already present in their organizations. They serve as the means for gaining insight into the underlying systems structures from which the archetypal behavior emerges. This is the most common use of the archetype.

Archetypes are effective tools for beginning to answer the question, "Why do we keep seeing the same problems recur over time?"

Prospectively, archetypes are useful for planning. As managers formulate the means by which they expect to accomplish their organizational ends, the archetypes can be applied to test whether policies and structures under consideration may be altering the organizational structure in such a manner as to produce the archetypal behavior. If managers find this to be the case, they can take remedial action before the changes are adopted and embedded in the organization's structure.

From my experience, archetypes can be highly effective when examining PMO and ITPO organizational structure.

The particular "systems archetype" described above is called "accidental adversaries" (see Figure 4.1) because it explains how groups of people who ought to be in partnership with each other and who want to be in partnership with each other (or at least state that they do) end up bitterly opposed. It applies to teams working across functions, to joint ventures between organizations, to union-management battles, to suppliers and manufacturers, to family disputes, and even to civil wars.

The classic case where this "accidental adversaries" structure was first articulated and recognized was a scenario involving Procter and Gamble (P&G) and Walmart. Both had the same goals—improving the effectiveness and profitability of their production/distribution system—but they each felt that the other was acting in self-serving ways that damaged the industry.

Walmart learned throughout the 1970s and 1980s that heavy discounting and price promotion of goods could boost market share, value, and improve profits. But price promotions created extra costs and difficulties for distributors

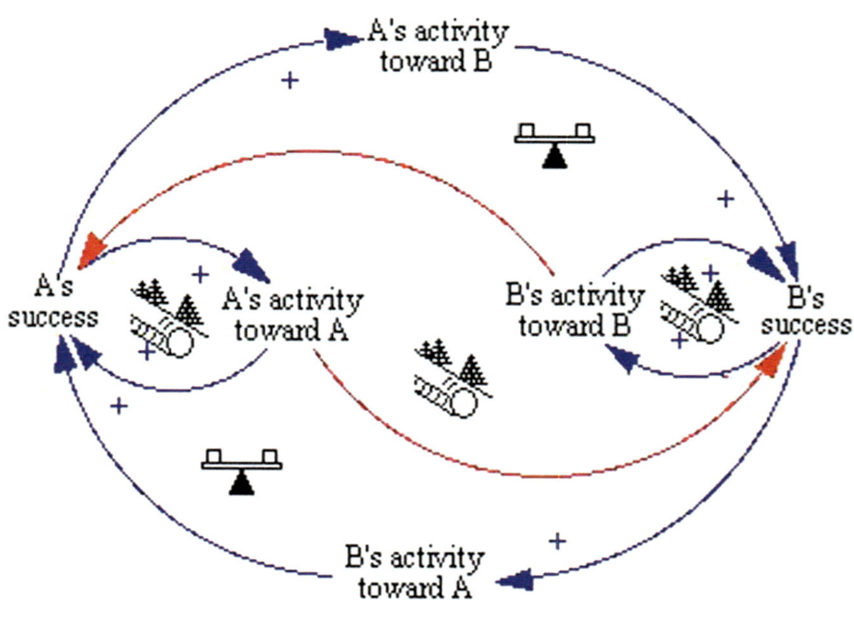

FIGURE 4.1
Accidental adversaries system archetype.

like P&G. The Walmart practice undermined P&G manufacturing, creating great swings in P&G's manufacturing volumes. The practices of each firm were intended to meet each firm's internal objectives but, as a partnership, each was always pointing a finger at the other claiming undermining practices. In responding more attentively to their internal objectives, the partnership fell short of optimizing its combined operational effectiveness.

While this pattern of behavior continued for years, attempts to reconcile and elaborate exactly what was happening was a difficult, if not impossible, order.

And so, this same pattern existed in the newly formed and maturing ITPO, among seemingly cooperative and optimization-oriented managers, who did not recognize the future implications of their antagonistic conduct. It is often not easy to discern or to admit that these behaviors take place among rational and intelligent groups who join their efforts to make a better condition within their groups.

But, as William Braun has suggested, the potential exists for archetypes to be applied to test whether the policies and structures under consideration may be altering the organizational structure in such a manner as to produce the archetypal behavior. If PMO managers find this to be the case, they can take remedial action before the changes are adopted and embedded in the organization's structure.

Have you identified some recurring behaviors in relationships between your key PMO suppliers, vendors, partners, or other support groups which can be limiting your attainment of PMO Excellence?

Here are some prescriptive actions and seven action steps from William Braun in case you find candidates for "Accidental Adversaries":

Prescriptive Action

- Revisit the original opportunity that brought the PMO parties together into a collaborative relationship.
- Use the archetype to identify the origins of adversarial attitudes.
- Renew the shared vision of the collaborative effort and commit to team learning.

Seven Action Steps

- Reconstruct the conditions that were the catalyst for collaboration and PMO success.

- Review the original understandings and expected mutual benefits.
- Identify conflicting incentives that may be driving adversarial behavior.
- Map the unintended side effects of each party's actions.
- Develop overarching PMO goals that align the efforts of the parties.
- Establish metrics to monitor collaborative behavior.
- Establish routine communication.

I don't want to leave you with the impression that this ITPO archetype produced a highly dysfunctional organization as it matured. Over a two- or three-year period, the management of the ITPO and the consulting firm realized and recognized the internal objectives of the other party, and actually began to discuss how they could support their own internal objectives while creating a fully functional ITPO that everyone envisioned at the beginning. Dialogue and continual realignment of values and vision were effective over time.

DOES STRUCTURE INFLUENCE BEHAVIOR? THE ROLE OF MENTAL MODELS IN DRIVING GROUP BEHAVIOR

In my book *Lessons Learned: Taking Project Management to a New Level in a Continuous Process Improvement Framework*, I talk at great length about the role that organizational dynamics plays in driving the behaviors and ultimately the performance of groups like PMOs or Project Offices.

Some people have asked me to elaborate more on this subject, so I want to provide a real-life example that vividly shows the relationship between mental models in groups/individual behavior.

When I worked for Ford Motor Company many years ago, I was a Product Planner in a non-automotive Product Planning group. The automotive Product Planning group was a group within Ford that drove product strategy and ultimately new product offerings for the company. Everyone considered automotive Product Planning to be the pinnacle of personal achievement and success within Ford. So, naturally, many people sought positions in automotive Product Planning because of the "stature" of the group.

One of my colleagues in the non-automotive Product Planning group decided he wanted to reach that pinnacle himself. So, about six months after joining our Product Planning group, he was able to transition to the automotive Product Planning group. Everything seemed rosy for a while. But then he noticed that the automotive Product Planning group had an unwritten rule that no one left the building until early morning … meaning that most of the group worked almost around the clock each day.

My friend began to think about how he could make a change in his own behavior. So he began to go to work earlier in the morning, thinking that arriving early meant that he could complete his day's work and leave for home before midnight each day.

Unfortunately, he found that the peer pressure and mental model that existed for the group members to be at work together, working diligently, until way past midnight every night was too ingrained to make a dent in.

What was happening here? Organizational dynamics teaches us that Vision, Mental Models, Systems Archetypes, Patterns of Behavior, and ultimately Events are the sequence of leveraging actions for scenarios like this one.

Events in this case were the daily schedules of the individual group members.

Patterns of behavior were what the group exhibited by being at work almost twenty-four hours every day.

The systems archetype from this example can best be summarized by the title: "I Can't Leave Work Until …"

But the mental model and the vision were really at play here.

Everyone in the company had a vision that the automotive Product Planning group was the "be all and end all" of elite groups. To be part of this group meant an identity within the company that was unsurpassed by even the president and CEO of Ford. The mental model that developed from this vision was one of personal diligence to uphold an image that "no one ever goes home."

How powerful can mental models be in an organization? Powerful enough to turn the most loyal employee into the product of the structure that induced the behavior. It is the same structure that the organization puts in place to drive results and outcomes and performance. But it also drives behaviors that are sometimes detrimental to overall performance.

In your project organization, are there mental models that need to be challenged and put to rest? Think about it!!!!

WHAT IS THE ACTUAL COST TO YOUR PMO FOR NOT CAPTURING AND SHARING PROJECT LESSONS LEARNED?

A few years ago, "Meet the Press" had an episode reflecting on the lessons learned from Hurricane Katrina, several years after the disaster had occurred.

The show stated that the disaster was the largest man-made disaster in U.S. history; not the largest "natural disaster." Why? Because at some point in the past, a large (failed) project was undertaken to reinforce New Orleans' levee system. This project was supposed to reinforce the levees so that they could withstand the flood waters of anticipated hurricanes and storms. Unfortunately, the levee system failed during Katrina.

A new levee system is now being installed by the Army Corps of Engineers. There are many lessons learned resulting from the previous levees' failure during Katrina, as well as many years' data from previous storms and from computer simulations. These lessons learned are currently being addressed by the design of the new levee system.

What is the actual cost for failure to capture and share Project Lessons Learned? Aside from the human cost of storms like Katrina, what is the actual project cost that is incurred by not capturing, documenting, sharing, and institutionalizing Project Lessons Learned? How can we get a manageable and actionable handle on the real contribution of Project Lessons Learned on saving future project costs?

The answer to this question obviously depends on many factors: the complexity of the project, the number of systems impacted, the number of dependencies of project info with other like systems, and so on. How can we get a handle on this type of information, and of what value would it be if we were to understand and apply it?

I once studied with a professor whose favorite expression was "Analysis is the Essence."

It did not really dawn on me what that statement meant until I encountered some business situations where bad decisions resulted in a failure to meet objectives such as "on time, on budget." In these cases, it seemed to me that the logical sequence of events should be "analysis" followed by "rational thought" followed by "decisions to proceed" followed by "actions."

We all know that while it is easy to recommend to others courses of action, or specific rationale, or well-thought-out research findings, it is very difficult for others to actually follow up and take the recommended courses of action.

Why is that? I believe that everyone has a tendency to believe that if they did not think of an idea themselves, then that idea is not of value to their ongoing, daily processes. And they may also be biased, and emotionally involved in the decision so that their "rationality" does not shine through. As is often said, you can clearly lead a horse to water, but you can't make him drink. It takes motivation and capability.

People decide to take courses of action based upon recommendations from others based on the credibility that they attach to the advice-giver and the usefulness of previous directions from that person. Leaders become leaders because they continually disallow their own thinking in favor of the more qualified thinking of their peers and associates; they have learned over time that their own thinking provides merely one perspective of a scenario that really demands many viewpoints to assess, understand, and take action upon.

So, how do you get people to embrace Project Lessons Learned—first, as a logical step in the project management process, and second, as a rational, thought-based process to provide information for future decisions about project work?

In the course of assisting project teams and PMOs with developing Project Lessons Learned, I have often encountered resistance to take the time to develop and capture lessons learned and to share this information with others. Emotional entanglement—as well as a lack of motivation in the sense that no immediate reward will be forthcoming (and perhaps that negative consequences may ensue)—often dictates the action or inaction. And I am sure, if you are actively engaged in project work in a PMO, that you have encountered the same.

So let me suggest another tactic.

There is a cost to be borne by the PMO for not capturing and sharing Project Lessons Learned.

To get a simple model for this cost, let's assume a model that is often used to get across the point that introducing changes at various key points in a project introduces additional costs to accommodate those changes.

If a project has, for example, four distinct phases, and if a change is introduced during the first phase that costs $10, that same change may cost $100 if introduced during the second phase, $1000 if introduced during the third phase, and $10,000 if introduced during the fourth phase.

Now, suppose that a project lesson learned is identified in phase four and that project lesson learned could impact a change at that point so that the original change in phase one would cost 0.001 times as much.

Suppose two or three Project Lessons Learned could be identified for every project. Thinking in an integrative manner to identify where projects could have been improved at previous project phases can deliver real cost savings. The point here that you should grasp is not whether the savings are 0.001 times the final cost of the change but that this perspective can yield significant savings to any project, no matter how complex or simple. Train yourself as a project manager to think in these terms and you will always be able to find significant opportunities in lessons learned.

Hard dollar reductions are the result of Project Lessons Learned.

What is the situation in your PMO? Are project managers willing to share Project Lessons Learned? Does the organization have a process for documenting and sharing Project Lessons Learned with future project teams?

HOW CAN THE PROCESS OF OSMOSIS POSITIVELY IMPACT PROJECT MANAGER PERFORMANCE?

It's funny how the brain and your mind work together—even when you are not engaged consciously in an activity.

Take for example one of the Lou Tice principles I have often referred to in this book.

Lou Tice often said that managers should not wait until they have assembled all their resources to make bold plans. Instead, managers should focus their attention and vision on bold plans first. Then the mind and bring will use their energy to find the solutions for the bold plans in both conscious and unconscious undertakings. The power of the Mind and brain combination is truly remarkable.

Now what does this have to do with the process of osmosis and project manager performance?

A while back, my wife and I were watching an episode of Trisha Yearwood's cooking show on one of the "Food Channel Networks." Trisha's theme was an homage to Julia Childs and her French cooking which Trisha's mother had introduced her to many years ago.

In this current episode, Trisha and her sister were preparing a beef dish that required seasoning as part of the cooking process. They were using a "sachet," which is a small porous cloth bag or pouch containing spices, herbs, or vegetables as the seasoning ingredients. As they described the role of the sachet, I was surprised by how much I knew about "sachets."

I honestly have never done enough cooking to have prepared and used a "sachet." Yet, as I listened to the TV program, I felt that somehow in my experience I had encountered "sachets" before. As I thought about this, I recalled that as a youth my family would often gather on holidays and large family meals were often a focal point.

I recalled running through the kitchen engaged in some activity with my numerous cousins as these meals were taking shape. In my memory, the word "sachet" was often a topic of the family "chefs."

So why was I so attuned to the role and makeup of a "sachet?" Somewhere, somehow along this journey that we call "growing up," my mind and brain were made aware of the "sachet." It was almost like osmosis, a process whereby materials kept isolated by a porous barrier often experience a mixing of the individual materials.

Every day we are exposed to a myriad of experiences, both consciously and unconsciously, that shape the way we see and react to the world around us. Our behavior and actions are the result of this exposure. As project managers, we all experience actions, decisions, and behavior of other project managers and team members and stakeholders. We are experiencing what worked and what did not work, what decisions and plans were fruitful, and which ones failed. We are experiencing what may be termed "subconscious benchmarking."

Our brains and minds have a vast amount of surrounding experiences to assess and choose from. It's like osmosis, but we must be sure the membrane is open and receptive to these experiences. Like the knowledge of the "sachet" in my example above, we have to give our minds and brains room to function.

Have you created the type of environment in your project management organization which promotes an awareness and appreciation of project experiences that can benefit individuals and the entire organization? Like the "sachet," you can season your organization with such thinking.

WHAT'S RESEARCH WITHOUT PROJECT MANAGEMENT?

Millions of people are engaged in research today. Some of them just want to satisfy a curiosity they have about a question or problem. Many are seeking to satisfy a need of society. Many are digging into a new technology solution which will add functionality to new or existing products or systems.

Research may often involve looking at old problems from new perspectives. Albert Einstein said:

> The mere formulation of a problem is far more essential than its solution which may be merely a matter of mathematical or experimental skill. To raise new questions, new possibilities, to regard old problems from a new angle requires creative imagination and makes real advances in science.

RESEARCH is a fundamental and basic need to meet the demands of modern life. Often research is a key to new innovation which drives our society forward and enables our ability to satisfy our collective needs.

Many who have employed RESEARCH for a time have learned that there is a research process that derives from their interests and their approach to the questions at hand. But the basic question they face is how to turn their research strategy into actionable results and solutions. That is where project management comes into play.

To convert a research question or problem to a project with a fixed start and end date and to devote or dedicate resources and effort to accomplish the tasks and activities that lead to tangible results is the essence of project management. And don't forget to consider a Risk Analysis to identify potential events that may doom your project to failure.

Any researcher can develop their project management skills to meet the needs of their RESEARCH. A resource I would highly recommend is a 2016 CRC text *Project Management for Research: A Guide for Graduate Students* by Badiru, Rusnock, and Valencia. I have conducted many research projects in my career and directed current research students in many fields.

RESEARCH will be our guide to the future. Embrace it and all its potential for our society.

WHAT IS SO SIGNIFICANT ABOUT "FRAMING" THE PROJECT CORRECTLY?

I have used the term "framing" with regard to projects many times in this book. Some of you may be wondering just what is the significance of "framing" to the success of projects.

When I was a member of the Shared Services PMO with ConocoPhillips, we had a framework for analysis known as Decision and Risk Analysis (DRA). This framework (decision process) was applicable to both operational and project initiatives in the corporation.

One of the earliest and most important steps in the DRA framework was framing. When framing, we asked, "What problem am I trying to solve with this initiative?"

For projects, it is very important to understand the answer to the question: "What problem am I trying to solve with this project?"

Answering the question means understanding the situation and business context well enough that you can determine when a problem has occurred in the corporation and then determine a course of action for solving the problem.

In order to understand the significance of proper framing, I would like to relate to you a situation from my own recent experience. As I was driving from point A to point B the other day, I was also listening to a call-in show on the radio. It involved a caller with a real-life problem and a telephone consultant answering the caller's question by providing a different perspective and suggesting various action plans that might alleviate the caller's concerns.

The caller told the consultant that he was despondent about a recent phone call that he had received. He had been married for about ten years, and he and his wife had several children together. He had recently received an anonymous phone call, however, in which he was told that four years ago, his wife had an affair. The man said he believed the anonymous caller because of the details that he gave him about his wife, their family, and about his in-laws.

The caller told the consultant that he had always believed that he had a good marriage and that his wife had always "acted" as if she wanted to continue in their marriage, with no indication that she had been unhappy. The caller was beside himself because he could not understand the situation. He said that after he received the call, he confronted his wife and she said that she did have a short affair four years ago and that it was long over. He confronted also his in-laws, who confirmed that they had known about the affair, but said that he would have to talk with his wife about any details. The caller felt betrayed, even though his wife had tried to reassure him of her love and her devotion to their family.

The consultant listened to this narrative very carefully and then said, "Sir, over the past four years, has your wife been devoted to your family as a loving wife?" He replied that she had. The consultant said, "Did it ever occur to you that she chose YOU, rather than the other person, to continue her life with? Did it ever occur to you that your in-laws were actually trying to protect you from the truth because they knew the affair was short-lived and that your wife had dedicated herself to your home? Did it ever occur to you that the anonymous call was intended to hurt you and your wife because the caller knew he had lost any ability to influence the situation?"

The caller thought for a minute and then remarked, "No, I never thought of it that way. I was so hurt by that anonymous call that I had to lash out at someone. And my wife was the closest thing to me."

"She chose you," said the consultant. "Go home and thank her for being the kind of wife she is and the woman you love."

Whether you agree with the consultant's advice, she had effectively reframed the caller's problem.

How often do project managers get so caught up in the heat of planning a project, that they do not take time to "frame" the project correctly? On the surface, they may see an immediate problem that the corporation is facing, and so they plan the project to alleviate that obvious problem. But others may have a different perspective and bring different viewpoints to the table. Examining the different viewpoints is essential to correctly "framing" the project.

Herb Cohen is a master negotiator and author of a number of books on negotiations. His main theme in negotiations is that "an effective negotiator must care but not too much." What he means is that, if a negotiator brings too much emotion to a problem or is tied too closely to the emotional outcome of a particular negotiation, he may lose his impartiality. He may incorrectly "frame" the negotiation. He may develop a bias toward one party or the other. He may render himself ineffective in the negotiation.

A project manager faces the same situation when "framing" a project. The next time you plan a project, keep this caller/consultant story in mind. Then sit back and ask yourself "Have I framed this project correctly? Have I really described the problem that I am trying to solve? Or is it being masked by my emotion or by some other circumstances in the organization that I must uncover?"

Success lies in getting the "ingredients" of the pudding right and the right pudding ingredients lie in the "framing."

PATTERNS OF BEHAVIOR AND PROJECT LESSONS LEARNED: INFLUENCE OF THE PROJECT ENVIRONMENT

I am interested in two cases in which Project Lessons Learned can be identified from projects. I call these the Single Project Case and the Multiple Project Case.

The **Single Project Case** involves a project manager and his team who wish to identify, document, and share Project Lessons Learned at the completion of a project, usually during the Project Close process.

The **Multiple Project Case** is the case in which several projects are subject to the same "project environment" and, therefore, may exhibit patterns of behavior that are the result of the structure of the project environment. The "structure" of the project environment is the policies, procedures, standards, and working processes that the project organization establishes to govern the way that the project will be conducted.

As we have discussed previously, the real leverage in looking at patterns of behavior from several projects that are subject to the same project environment is that a single change to the structure could potentially improve the performance of **all** the projects in the environment. This is because the structure has dictated some behavior or action on the part of project teams or others associated with the project that has contributed to suboptimized performance.

Here is a simple analogy to further explain the power and leverage associated with looking at patterns of behavior among projects and tracing these patterns to some action or process in the project environment that contributed to those patterns.

The simple analogy is the story of a breakfast buffet at a hotel. From 6:00 AM to 9:00 AM each day, the hotel offers a breakfast buffet to its guests consisting of the usual breakfast foods: eggs, sausage, oatmeal, coffee, orange juice, muffins, bagels, cereal, and so on. Guests usually serve themselves from a buffet and then sit in a lobby designed with tables for use by the guests during breakfasts and other meeting occasions. Guests usually serve themselves and then clean up at the end, which assists the breakfast staff in maintaining a continuous flow of breakfast guests.

On one particular morning, as each table completed its breakfast and vacated the area, a member of the breakfast staff wiped the table clean with a damp cloth in preparation for the next guests. Several breakfast staff noted that there was a sticky substance on each of the tables that resisted the usual wiping with a damp cloth so additional cleaning was required. The incident continued throughout the morning.

At about 8:50 AM, the manager, who had been informed by the breakfast staff of the sticky substance, observed several guests as they ate and completed their breakfasts. He approached several tables and asked if the breakfast staff could examine the tables while the cups, plates, and utensils were still on the table. It was found that the sticky substance had been deposited on the table by the plastic orange juice cups.

A "pattern of behavior" had been identified. A sticky substance had been identified with each table where an orange juice cup had been placed by a guest. The sticky substance appeared to be the same in each case.

The manager and several breakfast staff proceeded to the orange juice dispenser and observed the following process. Guests would pick up a juice cup from a stack of empty juice cups and place the cup on a specific spot in the juice dispenser. The guest would then push a button to dispense the juice into the juice cup.

The manager and the breakfast staff examined the single spot where each empty juice cup was placed by guests. On that spot, there was a sticky substance.

A process that had been put in place to facilitate ease of juice dispensing had caused a bad outcome at the customer environment location. The process supported the overall structure of breakfast for guests during the 6:00 AM to 9:00 AM time period.

If the breakfast staff had not acted to identify the incident and its resulting pattern of behavior, the impact on customer satisfaction might have been negative. At the very least, it might have contributed to the sticky substance being spread to other equipment and guest utensils.

Because the breakfast staff and the manager reacted quickly, however, the sticky substance was removed from the orange juice dispenser.

This analogy is far from being too simple. It points out that **"patterns of behavior" are so important in a project or customer environment that project groups should be attentive to them and act with haste to correct them.**

Many such patterns of behavior occur in project environments every day. Do you take the time to think through incidents and events in your own project environment which can be "patterns of behavior?"

The quick action of the breakfast staff and the manager in identifying the patterns of behavior was a significant step in making the improvement to the process which kept the structure working as planned.

In your own project environment, look for "patterns of behavior" and the underlying root causes for their occurrence. The leverage you can apply in your business context is enormous.

And the next time you drink a cup of orange juice, think of this example. You will be glad you did.

HOW CAN WE INFLUENCE THE PROJECT ENVIRONMENT TO POSITIVELY IMPACT ALL THE PROJECTS SUBJECT TO THAT ENVIRONMENT?

Project Lessons Learned can arise from two sources:

1. From the individual project in question, usually by the project team's analysis of their behaviors and actions leading to observed actual results.
2. From the project environment, usually by analysis of the "structure" of that environment and its effects on project team behavior.

Obviously, Project Lessons Learned come from post-project completion analysis. Project Lessons Learned are intended to inform and impact the future behavior of new and existing project managers in new future projects where similar project environments were considered the "as is" process state. By using some systemic thinking principles, it is possible to affect the behavior and results of many projects "playing in the same space" by making changes in the project environment. This is often termed as "leveraging actions" to the project portfolio.

Now what does that really mean?

The project environment is composed of the external corporate environment and the internally created corporate environment. Usually, the organizational and governance structures for a PMO group are defined and evolve over time as the organization evolves. The structure put in place by the organization has a great deal to do with how people act, behave, and make decisions within that project environment.

We have often heard the expression that "structure influences behavior." The structure of the project environment is made up of the policies, standards, procedures, defined relationships, reporting linkages, and so on that constitute the working environment within the firm. We are also aware that, in the dynamic complexity of project environments these days, well-intended actions can lead to unintended consequences. John Sterman at MIT has studied this type of behavior extensively. This is because the "cause and effect" are not close in space and time, and some nonlinearity may occur in which an action or decision on the part of a person or group may lead to unintended actions or behavior by others based upon their interpretations in their business context.

Organizations and groups have the "ability" to anticipate and plan what effects their PMO structures will have on project behaviors. Accordingly, they can take action over time to react and adapt to unintended results that are occurring as a result of the structure of the project environment. The project environment is a dynamic phenomenon; we can adjust the structure in order to influence project behavior. This approach to project management, however, is just in its infancy and requires some art along with the science.

For those of you who may have trouble visualizing how structure might influence behavior, let me offer an example from my experience. Many years ago, I worked as a Product Planner for a major domestic automotive manufacturer in the Detroit area. One of the benefits offered to top executives of the Division and the Company was a company automobile. Each morning when the executive arrived at the building, he was met by a service technician who asked if the executive had encountered any problems or had any service items that needed attention. At the end of the day, each executive drove away from the facility with a clean, serviced automobile. At one time, a certain service problem was identified in the field and reported through the service warranty network to the company. Reports and metrics summarizing the problem were reported through the ranks and eventually were highlighted to the executives in charge of product quality. However, none of the executives had encountered the problem—or if they had encountered it, the problem was fixed with "same day service." The executives' personal experiences led them to refuse to believe that the reports accurately identified the extent of the problem in the field. The reported problems were not taken seriously. In some cases, only one executive out of twenty might have encountered the problem and, if he weren't assigned to product quality, the problem was forgotten as just another "minor" flaw. As a result, the division failed to react to a mounting service problem. The structure of the executives' benefits—the company car—had influenced behavior to such an extent that there was actually denial that such a problem existed.

Let's examine another example a little closer to home in the PMO. In the early days of defining new PMO processes and procedures for ConocoPhillips, we recognized the need to define some "rules" for Project Justification and Approval. These rules called for the project teams to develop a business case to justify the project. As in similar situations with other PMOs, we attempted to define some structure to answer the question "How extensive does the business case analysis need to be and who should be reviewing it for final signoff as an approved project?" An early attempt to define this standard resulted in a "rule" that every proposed project over $1 million total cost must have a full business case with all economics and it must be signed off by the IT Authorization Board. All projects under $1 million total cost must define a partial business case with limited economics and must be signed off by a subset of the IT Authorization Board.

Now, what happened? You could probably guess that there were many projects proposed at $900,000 or two projects proposed at $800,000 and $850,000 that could easily have been combined into a more strategic and doable project. Remember: "Structure influences behavior." "Well-meant or intended actions can often lead to unintended consequences."

So, the bottom line for those of you who are designing new PMOs or are working with existing PMOs which have a set of policies, standards, and procedures in place to "guide" the planning and execution of projects is this: follow some simple evaluation questions I present here to lessen the risk that you will introduce some unintended consequences from your project environment:

1. Develop some scenarios to test out your policies, standards, processes, and procedures. Take a typical project and follow it through the Project Management Process to see what the behavior of project managers, team members, sponsors, or other stakeholders might be.
2. If your policies and processes have been in place for some length of time, test them to see if the thresholds for governance committee reviews are still valid. In other words, over time your PMO projects may have taken on different characteristics which have increased costs. The threshold values for review and signoff may no longer adequately cover the portfolio as you had originally planned.
3. If new technology is introduced rapidly into your process, have a review process in place to determine at the earliest point possible in the process if the new technologies are compatible with current technologies and infrastructure. Otherwise, you will very likely incur additional costs for these new technologies when they are evaluated.
4. Check the hurdle rates for your economic/financial analysis in your business cases at intervals to ensure that they are set consistently with the business objectives and the types of projects that you are deploying.
5. Survey your Project Authorization Governance Groups occasionally to make sure there is consistency in the way each member of the Group is viewing the mission and value proposition which projects present to the organization.

6. Examine key reporting relationships for the project team within the PMO and understand how the overall structure of the PMO influences where decisions are made and who makes them.
7. Identify and monitor any cultural or business context issues that might play a major role in the project environment for your organization or industry. Since these variables are dynamic and change over time, make sure you are evaluating all projects on a consistent basis. This is, of course, easier said than done. It involves being a student of the project environment and touching base with others in the PMO to ensure all bases are covered.
8. Continue to look at Project Lessons Learned and use the information to feed back to the front-end evaluation process we are describing at present.

There may always be some project environment variables that are elusive and for which you will not be able to easily identify unintended consequences of actions taken by the project manager, project team, sponsors, or other stakeholders. It does not mean that you should dismiss this analysis as having no value.

Remember: The more you know and understand about all the variables in the PMO and its organizational setting which can impact project team and stakeholder behavior, the quicker you can identify process improvements and achieve sustained project success. That is a sign of the MATURITY of your project process and is a goal worthy of striving for.

5

Project Lessons Learned

LOOKING BACKWARD, LOOKING FORWARD

I found it really interesting that people were all clamoring to get copies of Steve Jobs' biography. And the press, even now, looks to key events from his life. Why is this?

People are looking for that instant key to wisdom that Steve Jobs exhibited which made him a successful entrepreneur and innovator. People are not content to reflect on their own experiences about successful activities from their own lives and their key interactions and relationships with others in order to find the lessons learned that they desire.

Instead, they are looking for that instant wisdom that Steve Jobs talked about throughout his life. It is interesting that people are studying him now more than when he was alive.

One of my favorite quotes from Steve Jobs was the following:

> You can't connect the dots looking forward; you can only connect them looking backwards. So you have to trust that the dots will somehow connect in your future. You have to trust in something—your gut, destiny, life, karma, whatever. This approach has never let me down, and it has made all the difference in my life.

In the past, "Lessons Learned" evoked a connotation of taking time away from valuable activities that supported ongoing interests to look back at some activity already completed in order to record someone's mistakes and provide guidance to them on how to proceed next time.

DOI: 10.1201/9781003593645-5

We need to reframe our thinking about "Lessons Learned." We need to see "Lessons Learned" in any situation as an opportunity to improve performance, create more successful outcomes for ourselves and for the people affected by our activities, and improve the overall processes by which we live our daily lives.

In a recent audio CD interview for *Success* magazine, Dr. Oz stated an interesting perspective on getting second opinions for a diagnosis that physician might have given for an ailment or disease. Many people refuse to get second opinions because they fear that their physician will take offense at being second-guessed. But he stated a statistic that one of every three-second opinions results in a reversal of the original opinion. So a "reframing" of this situation means that if a second opinion were shared with the original physician with good, sound, medical, and scientific evidence that the second opinion should be acted upon, then the original physician stands to benefit as much from the opinion as does the patient. Think of how many patients the physician will see going forward for which he will have more knowledge and opinions to provide an original opinion. The "Lessons Learned" from this example are enormous in terms of future diagnoses and resulting treatments.

We must begin to think of "Lessons Learned" in terms of opportunity and the value that they add to all stakeholders including those in the future who could benefit from the capture, documentation, and sharing of the "Lessons Learned." To connect the dots in the future, as Steve Jobs alluded to in his quote, we need to use all of our backward looks to gain a better appreciation for what is possible in the future. We need to think of the value we can create for the world at large by creatively looking at lessons learned.

WHY IS "REFLECTION" SO IMPORTANT TO PROJECT LESSONS LEARNED?

By three methods we may learn wisdom:

—First, by reflection, which is noblest;
—Second, by imitation, which is easiest; and
—Third by experience, which is bitterest.

Confucius

Those of you who have seen the movie "The Ron Clark Story" already know about the remarkable efforts of a dedicated teacher in inner city New York who developed a learning atmosphere for his elementary students which contributed to them excelling in the classroom at the highest level in every subject. Subsequent to Ron Clark's success in the New York schools, he visited every state to talk with students, teachers, and school administrators about what he had learned and how his students performed.

What he found in his journeys were the same characteristics everywhere that he had identified to be success factors in his own classroom experiences. He found dedicated teachers and students and administrators everywhere who displayed some key attributes which, when tapped fully, created a learning experience that resulted in success and a learning environment that could not be denied regardless of the social setting or the prior experiences of the students and teachers.

When he wrote his second book *The Excellent Eleven*, he focused on eleven characteristics of high-performing classrooms, living and learning environments that contributed to student success. One of those characteristics was "Reflection." Ron used reflection to instill emotion in the students' experiences with their present-day studies. At the end of every major section of work, he had the students write down their impressions of what they had just experienced and how it affected them. What types of things were going on in their lives at the same time, what emotions they felt as they mastered each subject, and what relationships they experienced. Many of the students remarked that, after reading their own reflections a significant time later than the actual events, they were amazed at the images the reflections created in their minds and the tendency to revisit those images as they experienced new, more challenging situations and environments.

"Reflection" is something we don't usually allow ourselves to experience because we are too busy getting on to the next task at hand, too busy hurrying to the next assignment that we don't pay attention fully to what we have created in the last assignment, too busy grappling with the next hurdle because it is there rather than analyzing why we felt a certain way about our work just completed.

Yet reflection is the very essence of what many of Ron Clark's students stated as being the most significant experiences they remember and the most often called upon thoughts when they faced really formidable challenges in future years.

If you are a project manager or a project team member, make time to "reflect" in a written form at regular intervals what you feel about your experiences on projects, what observations you experienced when facing new hurdles, what paths of accomplishment you have just taken to reach a successful plateau, what thoughts you would share with another person who might be faced with a similar challenge.

The words of Confucius on Wisdom will resonate in your mind over and over as you attempt new and more challenging tasks. "Reflection" will help you to tackle all things that seem formidable. Sharing those reflections with others will help cement relationships that are so valuable to the maintenance of the community of project managers and team members going forward.

THE INFLUENCE OF LEARNING, SHARING, AND "LLEARNING" ON NEW PROCESS DEVELOPMENT

I have introduced a new term into the project management lexicon: LLearning.

"LLearning" means "learning from lessons of projects or operations in a group."

This is to be distinguished from Learning, which means "formal learning or programs introducing new skills or competencies."

Both are important to our discussion.

As project managers and PMO practitioners, we all recognize the importance of PROCESS, PROCESS DEVELOPMENT, and PROCESS IMPROVEMENT to the success of our organizational initiatives.

A PROCESS is a set of activities or tasks that, when performed in a specific sequence, yields an outcome or result versus an initial objective. PROCESS ensures consistency and repeatability of delivery of results and outcomes.

PROCESS DEVELOPMENT is essential to defining an EFFECTIVE and EFFICIENT PROCESS.

PROCESS IMPROVEMENT is essential to incorporate new lessons learned into the PROCESS.

Based on my experience working with project managers and PMOs, **I believe that several major, essential ingredients of PROCESS DEVELOPMENT are "Learning," "Sharing," and "LLearning."**

I would like to provide you with three examples of this contributing relationship from my background and research.

THE MAYO CLINIC

My wife had major knee replacement surgery at The Mayo Clinic in Phoenix, AZ. I recognized very quickly that The Mayo Clinic had developed a number of processes that I termed "well-defined processes."

A "well-defined process" is one that has known inputs, known outputs, and that has a very detailed process internally for converting the inputs to outputs.

The development of these processes to such a mature state obviously took many years and much effort on the part of many people to achieve. So, I began a journey to find out who was involved and how they did it.

The Mayo brothers, William and Charles, are generally regarded as the two individuals responsible for the development and success of The Mayo Clinic since its inception. Of course, they were aided greatly by their father (a doctor himself who was appointed Examining Physician for the Union troops in the Civil War by President Abraham Lincoln) who was instrumental in laying the groundwork in the 1800s for modernizing medicine in America.

The Mayo brothers' formal education, along with their collaboration with many surgeons, their embrace of new techniques from European medicine, and their sharing of what they learned with others, forged a new way of organizing hospitals and clinics in America.

Their visit to the Johns Hopkins Hospital and Medical School in 1894 also paved the way for further refinement of their processes. Working closely with William Henry Welch, often regarded as the Father of Modern American Medicine, they adopted many practices of the Johns Hopkins Hospital practitioners in their new Mayo Clinic.

"Sharing" was a cornerstone of their process development.

In fact, the very definition of "Clinic" in Mayo Clinic means "a collaborative and sharing atmosphere and environment where visiting surgeons come to learn from experts in the field."

The Mayo brothers excelled in new medical process development and research. It became the cornerstone for the ongoing success of The Mayo Clinic in medicine in America.

In the Foreword to a fantastic biography of the Mayo brothers, The Doctors Mayo, it states that "[i]n clinical research, Mayo has achieved a reputation for taking good ideas and perfecting their application in medicine and surgery."

Such is the essence of using Learning, Sharing, and LLearning in new process development.

The Mayo brothers were active in medicine at a time in American history when medicine was advancing rapidly from a poorly defined rudimentary discipline to modern techniques for surgery, the application of Joseph Lister's sterile surgical procedures, and understanding of internal body systems, functions, and diseases. Their ability to influence new processes by taking known advances and spreading their usage was noteworthy.

BP IN THE GULF OF MEXICO

BP was the source of the Deepwater Horizon oil spill in the Gulf of Mexico several years ago. Despite the efforts of many, the spill affected the Gulf Coast of the United States in many ways—economically, socially, politically, and so on.

Despite what many think today about BP's response and performance in the Gulf of Mexico, any company in BP's situation would necessarily learn an enormous amount about its internal processes and management, its technological capability, and its emergency response procedures, by going through such an experience.

Today the BP ads on the TV networks talk about their commitment to the Gulf and to the United States, their reliance on sharing experiences and learnings, and their ongoing technology development in well monitoring and safety.

Sharing and LLearning must certainly be part of BP's new process development.

PANAMA CANAL DEVELOPMENT AND EXPANSION PROGRAM

The history of the first and second Panama Canal projects, along with the current Expansion Program, is a story of the development of project and risk management techniques and practices over the past 150 years.

Elsewhere in this book I have detailed this development and highlighted the impact of lessons learned (LLearned) from project to project and the sharing of information and key new methods of analysis and process improvement that were used in this project.

I sincerely believe that "Learning," "Sharing," and "LLearned" played major roles in the advancement of both project and risk practices, and in the way project managers approached the huge project challenges they encountered in the Panama Canal.

What are some actions we can take within our own organizations to capitalize on "Learning," "Sharing," and "LLearning"?

1. Develop industry benchmarking
2. Develop internal Best Practices
3. Develop external Best Practices
4. Develop, document, and share Project Lessons Learned in a continuous process improvement framework
5. Understand our "own" organization's capabilities, competencies, skills, and needs.

Why should we as project managers and PMO practitioners need to be reminded of the roles of "Learning" and "Sharing" and "LLearned" in new process development?

Because we are seeking to improve and get better at our basic disciplines in life. We have chosen a "philosophy of life" as Louis Tice refers to it, which values order, discipline, and organizational tools for succeeding at turning strategy into action.

Project Management relies on PROCESS, PROCESS DEVELOPMENT, and PROCESS IMPROVEMENT.

It's your job to see that all these things work for you and your organization!!!

THE IMPORTANCE OF ACTIONABILITY AND ACCOUNTABILITY IN PERFORMANCE LESSONS LEARNED SCENARIOS

On a number of occasions, I have written about the need to go further with lessons learned than just capturing, documenting, and sharing meaningful lessons learned from projects or scenarios in which the actual results differ from the expected results of the scenario. By definition, lessons learned are "experiences" or "insights" which can and should be used to improve a process. That is what continuous improvement is all about.

Often it is the case that organizations do not carry through to ensure that the process improvement is finalized and the new process verified.

This is where accountability is so important. The organization must assign someone, an individual or a group, accountable for putting the process in place and verifying its "actionability."

Even organizations that have a strong lesson-learned culture are remiss in not following through with accountability. The organization suffers.

An example I have used in the past to illustrate this point is as follows:

Suppose there is a Quality Control (QC) check/step at a point in the process where critical measurements are taken on an essential output of the process. If the functionality of the final product does not meet the expected results, one reason may be a failure of the QC step. Lessons learned may result in the QC step being suspect. This could be an easy lesson learned to implement as a process improvement. Detecting it is key.

Lessons learned are all around us. Often outside groups combine lessons learned and best practices to assure that lasting change is the final and absolute result. Look for lessons learned wherever an actual performance result differs from an expected performance result.

LESSONS LEARNED FROM THE NOVEL COVID-19 PANDEMIC IN THE UNITED STATES 2020

As early as 2005, world leaders considered that a virus-type infection was likely to be created, either by bird-to-bird or bird-to-human interaction or by lab introduction.

In 2015, Bill Gates predicted that, based on his information, a virus would probably be identified in the United States within ten years.

The Obama Administration began to prepare for such a virus, by assembling information on likely viruses, preparing a tabletop exercise to practice a virus response, and identifying individuals with virus expertise as well as a list of certain essential products that might be useful to a team facing such a pandemic.

There is evidence that the COVID-19 virus was in the United States far earlier than initially believed. COVID-19 likely entered the West Coast of the United States in January 2020 through airline passengers from China. COVID-19 likely entered the East Coast of the United States in January 2020 through passengers arriving in New York from Europe.

Facts about the novel COVID-19 virus:

- Highly contagious, usually spread by airborne particles, although surfaces can contain the virus
- Infected individuals may be symptomatic or asymptomatic
- First organs of attack are the lungs and pulmonary system
- Usually infects the elderly with co-morbidity conditions (diabetes, heart disease)
- Can affect young children who develop inflammatory conditions
- Virus information changes daily
- The proximity of individuals is key to transmission, so keeping individuals separated ("socially distanced") is a key to controlling the virus.

What Is Project Management?

Project management is a framework and methodology designed to turn "strategy" into "action" for an individual or organization. Larger corporations and governments often employ project management to create new products and services, as well as to develop integrated plans to combat threats to the national well-being.

Project Management Lessons Learned

Project management lessons learned are insights or experiences that can be used to improve the project management process of the particular project at hand. There are two types of lessons learned: Performance-Based and Structure Influencing Behavior.

In the first, actual versus expected performance is measured for a project manager, and the gap between the two becomes a candidate for a lesson learned.

In the second, and sometimes a more important branch of Project Lessons Learned, the age-old question is "Does Structure Influence Organizational Behavior?" Structure in this context is the processes, standards, procedures, and management objectives that guide a modern organization. Sometimes, management objectives create behavior on the part of individuals, or the organization itself, which is detrimental to the overall health of the organization.

Such structures might include the following:

- controlling the space occupied by a specific organizational entity, thus controlling cost;
- designing facilities so that they occupy the minimum space, but contain essential highly trained expertise, thereby reducing costs;
- designing work structures that include very close work conditions;
- identifying subject matter areas of the organization, such as Security or Cyber Security, which can be used to contract specific expertise during certain project time.

In corporations, I have encountered organizations where structure influences behavior most often in specialized security or cyber groups, where expertise is required for a short time and can be contracted by the project manager so that the expertise is available when needed. Most often, when the expertise is needed by the project, the Security individual is not available, thus requiring the project manager to add additional cost to find a replacement.

Project Management Lessons Learned from the COVID Pandemic Crisis in the United States 2020

Project Management Performance–Actual versus Expected

THERE WAS NO PROJECT CHARTER FOR THIS PROJECT. Therefore, no stakeholders were named, and no sponsors were present to guide the invisible project team.

No assigned project manager was identified. No resources were allocated to the project, and no special expertise was identified.

The individual state governors, such as Andrew Cuomo of New York, had to take the lead as soon as the virus started appearing in their states.

Any actual performance was by the individual who encountered a deviation from norm.

Performance management results were "ad hoc."

These actions certainly caused the cost of the entire project to exceed that which might have been planned to be spent on the project.

No plans for testing, drug discovery, or planning were apparent.

Project Management Performance—Does Structure Influence Organizational Behavior

Here are some categories of structural issues in the 2020 Covid-19 pandemic crisis

- Prisons. Prisons are a case where many individuals are in close proximity due to incarceration. In the future, prisons could be relieved of their personnel count.
- Nursing Homes. Nursing Homes were one of the first highly hit pandemic areas. In the future, relieving Nursing Homes of their patient load could reduce the counts.
- Military Installations, including warships. The USS Theodore Roosevelt, for example, contained many individuals who could easily contract coronavirus because they were in close proximity. In the future, I would advise drydocking all ships and relieving all non-essential personnel.
- Food Handling Personnel. In many situations, food handling personnel are required to work in close proximity, as are manufacturing line workers. This created an internal crisis of how to handle workers infecting each other. This crisis often caused meat processors to shut down and manufacturing facilities to refine their assembly line processes.
- Hospitals. Highly skilled healthcare workers in hospitals continued their jobs day-to-day. This creates another identifiable group if another pandemic hits. These workers are on the front line every day and return to their homes each evening, thus creating more opportunities for infection.

There is no doubt that the COVID pandemic of 2020 cost the United States of America great financial loss, as well as the loss of over 170,000 lives at last count. The Lessons Learned are great in terms of the structure influencing organizational behavior. Since this pandemic is inevitably going to repeat itself, the United States must prepare or pay the price again.

LESSONS LEARNED DON'T HAVE TO IMPLY BIG PROCESS

Consider the fact that lessons learned are all around us in everyday life. How we choose to capitalize on their usefulness is our own choice. A quote that

I often hear is "If we do not learn from history, we are destined to repeat it." How true this statement is!!!!

Every process, large and small, can benefit from a logical, rational approach to identifying the lessons from the situation and then finding the appropriate place to improve the process with the lessons. In your own life, look for instances where what was expected differs from what actually happened.

Ask yourself the question "What caused this deviation?" If others are involved in the situation, you might have to talk with them to get their perspectives. Once you identify a change that can create the expected result or which is projected to create the expected result, you can implement the change and then monitor the outcome.

So find some small thing in one of your everyday processes that needs changing and change it. You will be surprised at how fast this leads to searching for larger processes you can change.

I have always felt that the people who are directly involved in a process are the best equipped to know how to find lasting improvements. But you can also resort to BENCHMARKING and BEST PRACTICES to find improvement candidates. Just remember, it doesn't have to be a BIG PROCESS to find some very significant improvements.

Lessons learned are all around us. They can be our best friend in making our lives happier and healthier. It just takes responding to the situation in a proactive way. You can accept this challenge.

LESSONS LEARNED ARE ALL AROUND US

The framework for lessons learned for projects has been very simple: What is the expected result; what was the actual result; what was the gap between the two, and what was the lesson learned based on this gap? In its simplest form, this is merely a fact-based investigative examination of a scenario for a project.

But it need not apply only to project scenarios. It can be applied to **any** scenario that is fact-based and needs a close examination of the perspectives and the deliverables from the scenario.

Project teams can use this fact-based investigative framework to examine their everyday workflows and processes where deliverables and performance expectations are the norms for activity. Believe it or not, most team activity that follows an "action" leads to "outcome," which actually results in unintended consequences due to some factor that the team did not take into consideration prior to their action. The only way to sort out the actual result is to use a fact-based investigative framework to see what influences the team did not take into consideration.

Take for example a procurement scenario where the project team is seeking an outsourced application for its work. Even a closely drafted contract for the procurement of this app can result in some functional requirements of the app being overlooked in the development. The delivered product might not match the business or functional requirements outlined in the project charter. These unintended consequences need unraveling to ensure the project team gets what it intended to procure.

The fact-based investigative framework presented in my book *Project Management Lessons Learned: A Continuous Process Improvement Framework* can help you sort out these scenarios with unintended consequences. Try it out.

THE IMPORTANCE OF PERCEPTION IN CAPTURING AND DOCUMENTING PROJECT LESSONS LEARNED

Every time you take an action with a result, there is an opportunity for lessons learned.

Expected results versus actual results is the key to defining lessons learned. But capturing, documenting, and sharing lessons learned add a new dimension to the equation. The old "After Action Review Framework" is just the starting point.

So, what are some of the more subtle points in developing truly actionable lessons learned?

1. As much as possible, try to capture lessons learned as quickly as possible after the action.
2. If your PMO has a Lessons Learned Framework in place, use it. Do not reinvent the wheel. It will reinforce continued use and result in consistent and repeatable outcomes.
3. Be prepared to reconcile differences in perception and perspective in different observers of the action and result.
4. If your organization does not have a Lessons Learned Framework, use your project team's experience to promote the adoption of lessons

learned—and the consistent use of a single framework—throughout the PMO. Your project team can add real value to promoting a consistent Lessons Learned culture.

5. Use an online repository to store and share lessons learned. Breakfast Forums are a good way to get project managers accustomed to publicly share experiences.
6. "Actionable" means being prepared to turn lessons learned into lasting process improvements for the organization. Remember: no process was ever created that could not be improved by judiciously applying lessons learned.

Lessons learned are the lifeblood of a mature PMO culture. A good lessons learned process will provide the headway for a successful organization.

A LITTLE MOOD MUSIC, MAESTRO!!!

As project managers, we have all participated in Project Reviews and Project Closeout Sessions that looked and felt more like "inquisitions" than positive, healthy, and constructive reviews. "What went wrong? Who is to blame? Who was responsible for these outcomes and results? These actual results were not expected and we are going to get to the bottom of the story."

There is a certain mood created by reviews such as these that perpetuate similar behavior in subsequent reviews. They take on a habitual characteristic. How does this happen and how might we as project management practitioners change these practices to improve our project management processes in a positive, sustainable, and constructive way?

Each of us has often found ourselves "whistling" or "humming" a tune which has stuck in our heads. You could be walking through a department or grocery store with some background music softly playing and suddenly you are humming or whistling it too. Or you could see a musical performance that really struck a "resonance" in your heart or mind.

We often say that this music sets the "mood" or "tone" for our behavior and actions. For example, I can recall as a young child hearing my mother softly singing:

- what a beautiful morning,
- what a beautiful day.
- what a beautiful feeling,

Everything's going my way.

I didn't think much about it at the time but, as I grew older, I began to recognize that what she was singing actually represented her philosophy or attitude and approach to life. The more she carried this tune with her throughout the day and week, the more her behavior and actions and those of the people around her mirrored that tone.

I would say that there is something infectious about the mood music you have in your head, so much so that I want to challenge each of you to find that mood music that improves your performance and that of your projects in every area.

Once you do that, you will realize that you personally can change the mood and tone of your Project Reviews and Closeouts to reflect your own philosophy of what a successful project should look and feel like.

For project managers who are familiar with the older ways of conducting project reviews, there exists an opportunity to change completely the "mood music" of these reviews. YOU CAN BREAK THE HABIT. Focusing on Project Lessons Learned in a positive manner would establish a "mood" or "tone" for all project managers in an organization or group which they could "whistle" or "hum" as they go about their daily project manager tasks.

Is this thinking too radical for the project community to embrace and use as a new standard for conducting actionable project reviews? I really think not. We have all been impacted by "mood music" in all facets of our lives, and this is just another application of that concept.

You as an active project manager in your organization can help ingrain this thinking and approach in your project practices. You can forge a new philosophy and approach to your project management career going forward by embracing these ideas.

In a June 2014 issue of *Scientific American* is an article "The Neuroscience of Habits." One of the major findings recently is that short-term decisions and judgments are often directed by habit rather than by rational and reasoned study of all the issues. "Mood music" works the same as habits. So if we can get project managers to embrace Project Lessons Learned as the most up-to-date "mood music," they will begin to use it in their habitual thinking going forward.

Before his death, noted leadership development teacher Louis Tice of The Pacific Institute often stated that "People act in accordance with the truth as they perceive it to be." What if that TRUTH were guided by mood music which valued lessons learned as a contributor to advancement of process and procedures? When I was a project coordinator with ConocoPhillips, we initiated some Project Lessons Learned reviews to give our project managers some experience in identifying, capturing, and sharing lessons learned from projects. We established a Microsoft SharePoint database for storing these lessons learned for future reference. In doing so, I would like to think that we started the "mood music" for project managers to follow going forward which focused less on who was to blame for project shortfalls and more on the improvement of the project process going forward for the entire organization.

So, what is the bottom line here? If you ensure that your project environment is providing the right background and mood music to your project managers, you can cement the habitual reactions that will lead to identifying, capturing, and sharing Project Lessons Learned for the continuous improvement of your business.

So, come on Maestro … a little mood music if you please!!!

WHAT NEW SKILLS AND CAPABILITIES MUST A PROJECT ORGANIZATION DEVELOP IN ITS PROJECT MANAGERS AND FACILITATORS TO ENSURE A ROBUST PROJECT LESSONS LEARNED PROCESS?

What are some of the prerequisites for an actionable Project Closeout and Lessons Learned Process for the PMO or project organization? In general, these have been activities or competencies or capabilities that the organization needs to nurture in order to really make Project Lessons Learned an active contributor to the overall project process.

In the course of this review of critical success competencies, we must also ask **what skills and capabilities are required of project managers or project facilitators to enable the full development of Project Lessons Learned?**

As we all know, people make a difference in the deployment of corporate resources. To make a real difference, and to stand out versus other organizations' implementations of the Project Lessons Learned process, your organization needs to ask what skills and capabilities it must develop and nurture in its project managers to fully realize the benefits of Project Closeout and Lessons Learned.

It is the combination of Process, People, Tools, and Capabilities that really make a successful package for Project Closeout and Lessons Learned.

So, what skills do your project managers need to develop to enable the full development of Project Lessons Learned?

First, the project manager must be a good facilitator of the perspectives, facts, and deliverables that are key to defining Significant Events for Lessons Learned. The project manager or coordinator must be able to work with different perspectives as to what actually happened in key project situations, sort through the various perspectives, and then "reconcile" if necessary those perspectives that would hinder gaining consensus on key Significant Events.

Second, the project manager or coordinator must be a good "reframer" if necessary to hold the dialogue with key project participants who might see situations differently but, in actuality, just need a "reframing" of the facts of the situation to be brought on board.

Third, the project manager or coordinator needs to be a leader of the discussion, at all levels, during which it is important to convert Significant Events, to Candidates for Lessons Learned, and finally, to the identification of Lessons Learned themselves.

Fourth, the project manager or coordinator needs to be a good documenter of the process using the Lessons Learned Template as a key documentation tool.

Fifth, the project manager or coordinator needs to be able to explain and put into context why certain lessons learned led to improvements in the process, and how those improvements were actually realized.

In my experience, even the most experienced project managers have trouble working with the framework at first and must develop some experience with using it effectively. In most cases, it is these capabilities or competency areas that are being developed while the project manager or coordinator grapples with the details of Project Closeout and Lessons Learned.

I have also found in my work with PMOs that oral presentations of Lessons Learned by project managers to their peer group is a very effective way of building these capabilities and also to build project leadership. In a major PMO with which I worked, those project managers who consistently stepped forward and volunteered to lead this discussion of Project Lessons Learned eventually became the leaders of the PMO. Why was that? Project managers who embrace change, who believe that risk management is a friend, and who want to be "curators" of the project knowledge in their own project communities are often the very ones who step forward and lead in the Project Closeout and Lessons Learned arena, as well as in their own PMO development and maturity.

So, please keep in mind that it is not enough to espouse good Project Lessons Learned documentation. You must first look at how your organization is nurturing those Process, People, Capabilities, and Tools for Project Closeout and Lessons Learned. You can start this journey in your own organization by volunteering to speak to other project managers about your own project experiences and by beginning to examine Facts, Perspectives, and Deliverables for your recent projects in anticipation of capturing Lessons Learned. Integrate this zeal with a good Project Risk Management Plan and see how it contributes to the success of your PMO or project organization in upgrading the entire project process.

WHEN IS THE BEST TIME TO DOCUMENT THE PROJECT LESSONS LEARNED DURING YOUR PROJECT?

Note that the question above says the "best time" and not the "most convenient" time.

The "best time" should be that time which will allow for reflection, feedback, and adjustment of any project assumptions or other parameters that would contribute to project success. It should be that time which will allow win-win solutions to be identified from issues with different viewpoints, as opposed to the compromise solutions which often occur in projects when a full look back and "analysis" has not been conducted.

In the past, Project Lessons Learned was often modeled after the "After Action Review (AAR)" from the military, in which a review session would be conducted after a military operation in order to capture lessons learned and positively impact the behavior of the team in future exercises. This was a convenient time for such reviews.

The "AAR" methodology was carried over to modern project management much as other practices are carried over from one discipline to another without proper regard to what contributes the most to success in the future endeavors or in the current project at hand.

Instead of waiting until the end of your project, I suggest that you select intervals during the project for such reviews. These could correspond to "stage gates" if your project is organized in such a way, or they could occur at natural or major breakpoints in activities in the project schedule.

Documenting lessons learned at multiple points during a project allows a look back at the assumptions versus reality of the project and affords the opportunity to define new assumptions for the remaining portions of the project. Also, it allows time for "reframing" or asking the question "What problem are we trying to solve?" with this project.

How many projects have you seen completed where the project team, in looking back at what was accomplished, realized that they had addressed or answered the wrong question? "Framing" and "reframing" the major issues in the project are facilitated by conducting Project Lessons Learned at intervals throughout the project.

Now that you've decided to conduct your project review at intervals throughout the project, the next question becomes "What framework should I use for conducting a Project Lessons Learned session?"

The usual questions asked in a Project Lessons Learned exercise are as follows:

1. What was the expected result?
2. What was the actual result?
3. What is the deviation of actual from expected?
4. What is the lesson learned?

While this is a good starting point for capturing and documenting lessons learned, let's take this one step further.

I would recommend you look at the work of Roger Martin, Dean of The Rotman School of Management at The University of Toronto, to find a good framework for examining Project Lessons Learned during the course of the project.

Over the course of about fifteen years, Roger Martin studied how successful leaders think. Leaders from many different fields were interviewed and a thought pattern was discerned. Roger Martin termed this "Integrative Thinking" because it differs from "Conventional Thinking."

Among the features of "Integrative Thinking" which makes it a good methodology for examining Project Lessons Learned are the following:

1. Examination of more salient features of the project which might not have been considered when the original project was "framed."
2. An integrative look at other issues which might impact the project so that a holistic view of the project and its impact on all stakeholders and the environment is considered.
3. An attention to architecture of the project and its major issues which give a good overall view of the systems dynamics involved in the project.

This also provides clues to "dependencies" between your project's deliverables and those of other projects, that is, the fact that projects "give" and "get" information and deliverables during the course of the project life.

4. A consideration of opposing viewpoints which does not acquiesce or dominate other viewpoints but which takes into account the valuable features which contribute to a win-win situation rather than a compromise situation.

Let's take a simple example now.

Suppose your project has four distinct phases, each succeeding phase beginning at the completion of the previous phase in a linear fashion. Also, suppose you want to conduct a "Project Lessons Learned" session at the end of phase one.

The sequence of activities you might follow in conducting this lessons learned exercise might be as follows:

1. Look at results from phase one. Are there any assumptions that require revision prior to going into phase two based on the results of phase one? Have you correctly "framed" the problem in this project based on the results from phase one? Are there any other salient features that should have been considered in the project during phase one that were not considered?
2. From a holistic standpoint and anticipating activities in phase two, have we learned anything in phase one which would impact how we handle phase two (or for that matter, any later phase) activities?
3. Have we correctly planned the architecture of the project so that "cause" and "effect" of decisions and actions taken in phase one can be clearly discerned in phase two and later phases? Are we certain of the "dependencies" between projects and the key "gives" and "gets" between or among projects?
4. How are opposing viewpoints concerning decisions and actions in phase one being addressed? Are we taking the best features of opposing viewpoints to find the final solution or do we still seek compromise when opposing viewpoints clash? How will we proceed to a win-win solution for the project outcomes so that stakeholders are satisfied with the project outcomes?

In conclusion, look for opportunities to conduct Project Lessons Learned sessions and to document the lessons when it makes sense to conduct them.

The real value of project management is when lessons learned are fed back into project schedules and plans and positively impact project team behavior and resulting project decisions. The success of projects is more assured if lessons learned are taken seriously and every team member takes ownership of both the identification and sharing of Project Lessons Learned.

PROJECT LESSONS LEARNED: WHERE IN THE WORLD HAS MEL BOST BEEN?

Why is *Success* magazine so successful?

Is it because the articles are so skillfully written that readers feel compelled to read the magazine cover to cover?

Is it because the insights are so insightful that no reader can resist the articles?

Is it because John C. Maxwell always seems to grace their pages with his deep insights into leadership?

No.

It's because everybody, no matter what their chosen field, or discipline, or career path, wants to "get better" and to "improve" their performance and their happiness in their chosen field. People are searching every day for that bit of wisdom that will give them a clue about their own lives and their own happiness.

If you are a seasoned project manager, an aspiring young project manager, or a practitioner from another discipline or field, the idea that something "learned" may contribute to your achievement and happiness is important.

And "Project Lessons Learned" is even more important because everyone takes on "projects" large and small, formal and informal, approved and unapproved, budgeted and unbudgeted, every single day.

I have been studying and writing about Project Lessons Learned for a number of years. I helped a major Fortune 500 company develop a robust Project Lessons Learned process and framework for their PMO organization. I have contributed to several podcasts on the subject that have been well received by the project community.

Dan Pink's work on *Drive* is also very relevant here. In studying groups that are engaged in highly complex work with significant intellectual content, he has found that three major motivators are at work:

1. **Autonomy**: This is the need for self-direction, the need to determine what and when a person will pursue in order to reach his own goals and objectives.
2. **Mastery**: People want to master their chosen "discipline" or field. In the case of project managers, mastery means project managers want to pursue Project Lessons Learned as the culmination of a successful project, as the pinnacle of sharing experiences and insights with the project community.
3. **Purpose**: People engaged in highly complex and intellectual work seek a higher purpose than just the profit motive for their efforts. They want to know that the larger community of which they are a part will benefit from their efforts. In the case of Project Lessons Learned, they want to know that the project community will benefit long term from the capture, documentation, and sharing of lessons learned.

Early in 2011, I predicted that we would see more PMOs focusing on Project Lessons Learned as a primary rather than a secondary focus (as has been the case in the recent past).

My interactions with many PMOs have revealed that this prediction is becoming true—more organizations are seeking to close out projects in a more formal, systematic, and documented manner. Project Lessons Learned is an excellent framework to follow when closing-out projects.

There is no doubt that those organizations who successfully convert Project Lessons Learned into process improvements will gain a competitive advantage.

6

Project Communications

WHY COMMUNICATION IS THE MOST IMPORTANT COMPETENCY IN PROJECT MANAGEMENT

As project managers, we have all studied the Project Management Body of Knowledge (PMBOK). This compilation is a summary of Best Practices and high-level definition of processes in planning and executing a project. We often think that developing business requirements, vendor management, and risk management are among those that require the most effort and compliance.

But, when you consider the importance of communication in the entire project management methodology and framework, it stands out as the link that joins the other competencies, the link that assures that we are considering all the inputs—the sponsor, the stakeholders, the developers, the project team, the suppliers, and the customers.

Communication is the most important competency that makes up the project management discipline. However, many people think that communication means that the project manager prepares the plans, communicates one way to his project team and suppliers, and watches the action. Nothing could be farther from the truth.

Communication is two ways, and sometimes three ways, in its linkage of team members with project manager and with suppliers.

Communication includes verbal, written, and postural components. Postural means the way we engage the listener or group, standing straight, taking charge of the message, recognizing others who want to contribute,

DOI: 10.1201/9781003593645-6

and engaging in discussion if there are different perspectives presented. Some think that postural communication is more important than verbal and written communication because it can express the hidden agendas of the speakers.

Why does communication matter so much?

- Communication through a project charter or other authorizing document is the official beginning or start of a project. It signals a willingness to proceed.
- Communication through project plans and schedules and resource allocation sets the "vision" for all the stakeholders on what is about to unfold and what it should look like at different stages of the project.
- Communication provides a "lesson learned" through feedback on the actual versus expected performance of the project process.
- Communication identifies perspectives on how the project should be conducted, differences of opinion, what resources it needs, and resolution of conflicts.
- The flow of information from project team members to the project manager is so important. Stories are the essence of learning.
- Communication identifies any organizational systems archetypes that need resolution such as "accidental adversaries," "limits to growth," or "tragedy of the commons."
- Communication supports the flow of information to support the "dependencies" competency, which identifies input deliverables to and from other projects. This is particularly important where projects conducted in parallel support each other with key deliverables.
- Communication supports Risk Management Plans by alerting the project team that a risk may have been triggered and needs a response.
- Communication formally ends a project and conveys to everyone whether any aspects of the planned project were not completed as planned or deferred to another plan.
- Communication of all aspects of human resources competency is essential for project governance.

The project manager must present a genuine authentic face to all stakeholders. Communication is the mechanism for conveying this authenticity. If the project manager has any conflicts of interest with the project in any way, he needs to reveal them as soon as possible and preferably before the project begins. Thanks for your attention!

CAN YOU HEAR ME NOW? WHAT ARE SOME EFFECTIVE COMMUNICATION MECHANISMS PROJECT MANAGERS CAN EMPLOY WITH TEAMS AND STAKEHOLDERS? PART I

One of my themes has been effective communication with ***all*** project stakeholders. Of course, this means using different types of communication, with different types of stakeholders, in different types of situations.

It has often been said that, if you want someone to talk with you in a meaningful way and to listen to what you have to say, you first have to say something of importance and value to the other person so that he will feel that you value his "stake" and continued involvement in the conversation.

Communication has several components. The simplest characterization is that something is conveyed by one party to another, and the "something" is received and understood by the other party. Without this sending and receiving, we do not have effective communication.

Four communication mechanisms are

1. Situational communication
2. Empathic communication
3. Resonance communication
4. Metaphorical communication

While you won't find these discussed in any standard project management texts or in the PMBOK section on "Communication," these four approaches to communication have been cropping up in my experiences with successful projects time and again. They seem to dominate the landscape of "real" project management. And in my experience, those project managers who are able to master these four methods of communication have flourished.

Situational Communication

Ken Blanchard's consulting group made famous a discipline known as "situational leadership." In essence, it meant defining the types of situations in which a leader finds himself and then adjusting leadership "styles" to the task at hand.

Communication in projects is much the same way. The project manager must recognize and be constantly aware of the situation and business context facing the project team and then adjust his leadership and communication approach accordingly.

In graduate school at the University of Michigan, Ann Arbor, studying nuclear science and engineering, I had a chance to work with a famous researcher, Dr. Ziya Akcasu, in his discipline of nuclear reactor control theory. Dr. Akcasu and I developed a research paper which I presented at the American Nuclear Society (ANS) Annual Meeting in Boston in June 1971. While I was attending the Boston meeting, Dr. Akcasu was on the West Coast attending another meeting.

On the morning of my presentation, I was second on the agenda to speak. As I entered the lecture room, I noted that there were about one hundred people in attendance, which was a good-sized group for the topics we would be covering. After the first speaker finished and answered questions, I walked to the podium and introduced myself and the paper topic. Most of the one hundred patrons had maintained their seats and, in fact, a few others had entered and lined the back walls of the lecture room.

Before I had a chance to continue, from the back of the room, someone asked, "Will Professor Akcasu be presenting today?" Although I was a little taken aback by the question, I replied that Dr. Akcasu was on the West Coast attending another conference and that I ("the designated graduate student and co-author") would be presenting our results.

Suddenly at least three-quarters of the room rose from their seats and left somewhat boisterously. After the dust settled, the remaining twenty-five or so people looked at me, and I looked at them. Then, without really thinking about what I was saying, these words came forth, "Well, now we have a smaller group who is really interested in the results of this groundbreaking work. We should be able to have a good open discussion of the details."

As people left the room that day, they thanked me for the intimate atmosphere I had created. I have often thought about what the lesson learned from that presentation might be. My answer: "Fame ensures millions of patrons but only the true scholars command the dedicated few."

Of course, my methods of communicating with the twenty-five or so remaining observers were quite different from the methods I might have employed with an audience of one hundred. This was a good example of situational communication in which a person adjusts his style to accommodate the change in situation, circumstance, or context. Pure lecture was certainly not called for with this intimate group.

I recall a Project Lessons Learned Breakfast Forum that I facilitated in my PMO work with ConocoPhillips. A very experienced financial systems project manager told the group that his greatest challenge in completing a major SAP ERP Financials project on time was that the project was dependent upon several other projects that were scheduled to be completed just before his project's deadline. In order to meet his own deadline, he spent a good deal of time in the last days of his own project ensuring that the other project managers successfully managed their own projects so that his project would not suffer any delays.

Such an endeavor is hard to delegate to others. It takes foresight and experience that cannot be easily conveyed to others. His management—and on-time completion—of the SAP ERP Financials project was another form of situational communication.

Empathic Communication

At the University of Michigan, Ann Arbor, I was a PhD candidate in Nuclear Engineering. One requirement for obtaining the PhD was passing a written and oral preliminary exam (called "prelims"), after which I could proceed to the research phase of my candidacy.

I had heard numerous worrisome reports from students who had failed the prelims three or four times before finally passing.

The prospect of failing the prelims terrified me, so I developed a strategy that defined two possible courses of action:

(1) I could spend all my waking hours studying every resource on nuclear science and engineering that I could get my hands on; or
(2) I could align myself with a professor who was doing research in a field of my interest and go into the prelims with a certain amount of research already behind me along with the endorsement of that professor.

I chose option two.

On the day of the oral examination, I had already taken the written portion of the prelims. While the written examination had been particularly difficult, I thought that I had done well and had a chance of passing the prelims on my first attempt.

At the oral exam, I nervously answered the questions (in a way that I thought was somewhat incomplete). After answering the questions, I then had a chance to describe some of the research on which I was working. At the end of the oral exam, I asked the assembled group of professors if I could say a word to them before leaving. I told the group that while I knew my opinion didn't really count, I did not think I had done as well as I had expected and that I knew I understood the field of Nuclear Engineering much better than had been conveyed during the oral exam.

The professors listened and then one well-known professor said "Mel, just remember this: No one has ever passed this test who should have failed, and no one has ever failed this test who should have passed."

To this day, I am uncertain exactly what he meant, but since I did end up passing the prelims, I think that it was his way of communicating with me that there was nothing else that I could have done in that oral exam that would have mattered to the final outcome. He was sensitive to how helpless the prelim exam process had made me feel, and he wanted me to know that, whatever the outcome, I had done everything I could do to prepare for the exam.

This was an example of empathic communication in which one party is sensitive to the feelings and explicit or implicit needs of the other party.

Shortly after the merger of Conoco and Phillips Petroleum in 2002, the newly merged corporation was faced with a dilemma—in some geographic regions of the world, there was operational overlap. One such area was Canada, where there were three or four distinct organizations with similar oil and gas operations and unique internal operational systems.

The decision was made to consolidate these operations into one Canadian organization going forward, with SAP being the ERP standard for all systems. This meant that business processes would be changing, and as a result, jobs, roles, and the people occupying those roles would also be changing.

A huge Organizational Change Management Plan (OCM) was implemented. When they chose the project manager for the project, it was someone who had recently completed a similar project in Europe. For the first two to three months of the project, the project manager and project sponsor traveled to all of the major operational centers of the organizations that were being consolidated. They held personal conversations with each employee to gain their input and to understand their current roles. These communications were extremely empathic in nature because not only did they seek each employee's buy-in, but they also conveyed to each that they would play a very important ongoing role in the transition from the old organization to the new. Some employees would become "power users" of the new systems, some would become casual users, and some would play a support role in providing key information for the systems. In other words, the success of the project, and that of the new consolidated organization, was in the hands of the employees and not in the hands of the project managers and sponsors.

Is that how it is in your organization? Do the users of the systems and the "supporting cast" feel like they have a real "stake" in the delivery and the outcomes of the key projects that will "make or break" the organization in the future?

Resonance Communication

In the sciences, we study a phenomenon called "resonance." Resonance is the state of a system in which an abnormally large response is produced by the system, as compared to a standard external stimulus. Once again, it is we are most interested in the "cause" and "effect" relationship. Resonances are observed in many physical systems. In nuclear physics, resonance occurs when there is a high probability of a certain type of action or reaction occurring if the conditions are right. For example, if a neutron of a certain energy comes in close range to a nucleus which has a high resonance cross section for absorption at that energy, then there is a high probability of absorption of that neutron by that atom.

There is an analogous situation in behavioral or emotional systems in which an external stimulus causes some form of a larger-than-expected response on the part of the person affected by the stimulus. Certain music, for example, is known to cause an emotional reaction. For example, the songs of Paul McCartney or Sting cause a "resonance" in the emotional behavior of many of their fans. This emotional response can be built upon to form a market for that particular music. The same can be said of certain foods. Pizza or Mexican food can, for example, cause a "resonance" response in some people who have a "taste" for that type of cuisine.

Perhaps you have never thought of markets in this way before. How can you estimate the market size for a new song by a particular group?

Now, how can an understanding of "resonance" help project managers communicate with project teams and stakeholders?

When I worked in a PMO for ConocoPhillips, we developed a number of SAP ERP projects for the organization during the merger of Conoco and Phillips Petroleum. After the first three or four of these projects were completed, we noticed that one aspect of the project which created longer actuals in terms of both budget and schedule was a phase known as "data conversion." "Data conversion" was a very general term that often included such competencies as "data scrubbing," "data interpretation," "data cleansing," as well as other related activities. We asked ourselves the question "Why does data conversion create such problems for project teams?"

What we found was that project teams, no matter how experienced in SAP projects, inevitably underestimate what it will take to convert the data from the original system into fully compatible and useable data for the SAP application. Typically, the number of resources would be underestimated, or the point in the project where data conversion would have to be initiated was miscalculated, or the "data conversion" team might be missing some key component of expertise.

How could we correct this situation and learn from past project mistakes in such a way that future project managers on SAP projects could fully anticipate the scope and resources required for an on-budget, on-schedule completion of "data conversion"?

We wrote a Project Lessons Learned document that would strike a "resonance" in any qualified project team. The project manager only had to discuss the Project Lesson with his team and his project planners in order to have a full and complete understanding of the problem, so that the team could understand and successfully act during the project's execution. By covering all the topics of data conversion, timing, competencies, special expertise, and inclusion of all known types of data shortcomings, the project manager was able to address any question that the project team might have regarding data.

Striking this "resonance" was essential to future on-budget, on-schedule performance in the SAP ERP arena.

So, as a forward-thinking project manager, look for those topics that will strike a "resonance" in each of your stakeholders, topics that you can call upon time and again to elicit a greater-than-expected response to your message. You will find that these topics are numerous and can be counted upon to reach project team members, sponsors, and steering committee members of business functional groups. Some of these topics might be OCM plans, risk analysis, and vendor management.

Metaphorical Communication

Metaphors are an effective way to put a situation or an issue into a context that another person or group can better understand. A metaphor is a figure of speech, symbol, or image that can help illustrate a point and connect similar or dissimilar items.

When people are able to see how to apply familiar processes and behaviors and attach their prior knowledge to various situations, it makes it easier for them to come up with action plans. It also enhances esteem and gives them confidence.

A number of years ago, after very successful runs in London and New York, Phantom of the Opera came to one of the performing arts centers of southern California. My family was very excited but, when I went to the box office to buy tickets, the earliest tickets I could get were for a performance four months in the future.

I had heard so many great things about the music of Andrew Lloyd Webber and the performances of his leading actors Michael Crawford and Sarah Brighton. So, I thought I would prepare myself for the upcoming performance by listening to the soundtrack on cassette tape. I had an expectation that I would be able to enjoy the atmosphere of the performance by listening to the music and the interaction of the characters on the soundtrack.

What happened? Without seeing how the performers interacted or how their roles were related, I really could not understand the context of the music or how it enabled the action. My understanding of the entire performance was incomplete. Many key things were missing. It was not until I saw the production with action, music, interaction, themes, and a plot that I realized what a fine tapestry that Andrew Lloyd Webber had woven.

Such is often the case with project teams. They may not see the full context of a project. Someone must communicate to them the "big picture"—the subtleties of the interaction among their roles communicated messages and feedback about the project, various decisions that had been made, and so on. That is where metaphors can be so valuable and effective.

On the Sunday, March 14, 2010, episode of "Meet the Press," Tom Friedman of *The New York Times* and Tom Brokaw, the guest moderator, were discussing the United States' role in the Middle East and the perception of the United States in the eyes of many of its residents. This exchange ensued:

MR. FRIEDMAN: Yeah. I mean, look, everyone in the Middle East is watching. You know, Tom, we both grew up in the Midwest. You remember, we used to talk about the Minnesota State Fair.

MR. BROKAW: Right.

MR. FRIEDMAN: I used to go to the state fair as a kid. There was a guy at the Minnesota State Fair who could guess your weight. I was fascinated with that as a five, ten-year-old. How does he get—and if he didn't get it right, you won a Kewpie doll or whatever. In the Middle East, people can guess your power from a hundred paces. They have to. That's how, how they survive. And if we look weak, vis-a-vis our closest ally in the region, that will have regional implications.

Tom Friedman used a terrific metaphor at that spot in the interview. Each television viewer was given a picture, indelibly etched in their mind, of a man in the state fair sizing up the power of someone he barely knows. And all for the sake of survival!!!

If you are a project manager, the next time you send one of your project team members to talk to someone in the business functional area about a potential change in project scope that the business functional group may be considering, why don't you use a metaphor when suggesting to your team member how the interaction should go? For example, if your project team member is a baseball fan, consider this approach:

"During this interview, why don't you approach Jim as if he is the starting pitcher in a baseball game, and he has loaded the bases in the bottom of the ninth inning, with your visiting team ahead 6 to 5 with two outs? It's crunch time! Let him know that you understand that he is upset for letting those batters get on base. But be forceful, and let him know that, if he allows a run or two to score, it's a whole new ballgame with your project. More scope means more dollars and inevitably missing the Job #1 date. Which is it going to be? Let him know that the project team and the sponsor would rather that he pitch a strike out, and that, together, you can still achieve that goal and win the game. If you frame it in such terms, he will deliver his best, and, more than likely, you both will emerge with a win-win."

7

Project Risk Management

THE DEVELOPMENT OF PROJECT MANAGEMENT AND PROJECT RISK MANAGEMENT PRACTICES: A REAL WORLD EXAMPLE—THE PANAMA CANAL

As I have mentioned previously in this book, in summer 2011 I traveled to Panama to facilitate two courses in Project Closeout and Lessons Learned for the Panama Canal Construction Division. The Panama Canal Authority is engaged in a Canal Expansion Program which is scheduled to complete in August 2014 which happens to be the Centennial of the opening of the original Panama Canal in 1914. This Expansion Program will not only allow larger ships to navigate the canal using a third set of locks but will make infrastructure improvements in water usage and environmental usage of land.

As far back as 1850, many countries around the world talked about the desirability of connecting the Atlantic and Pacific Oceans for transportation and navigation purposes. It was assumed that the most likely spot would be in the Isthmus of Panama which represents the narrowest land area bordering both the Atlantic and Pacific Oceans. That was probably the first of many assumptions that would be made about the Panama Canal and how it might be constructed. As most of you project managers know, "assumptions" can be the downfall of good projects and of those not planned or executed so well.

Little did I know in 2011 when I spent ten days in Panama that it would launch a new understanding of the advancement of both project management and project risk management practices.

In the 1850s, the desire was so great to create a transportation link across Panama from the Atlantic to the Pacific Oceans that a Panama Canal Railroad

DOI: 10.1201/9781003593645-7

was completed in 1855. Based on well-known railroad technology that had been utilized in many countries to that point in time, it set in motion a logistics and transportation evolution in Panama that has made that country a center for supply chain and logistics practices. Today, there are many seaports in Panama on both the Atlantic and Pacific sides, and the railroad links to all these ports have furthered this logistics development.

When I was in Panama in 2011, my taxi ride from the hotel to the Panama Canal Education Center where I facilitated my courses took me by many railroads and supply centers. It was really impressive. You may want to read more about Panama's Supply Chain and Logistics hub activities.

The history of the Panama Canal development since 1850 mirrors the birth, development, and improvement in project/risk management practices. And, as they say, "the proof is in the pudding."

For those of you who like to understand the background behind the development of a process or an invention or even a methodology, this is a real opportunity. I would recommend that you read Tom Kendrick's book *Identifying and Managing Project Risk: Essential Tools for Failure-Proofing Your Projects*. Each chapter of this book contains a section on some aspect of project/risk management as it applies in the case of the Panama Canal development. From the first unsuccessful project to the second successful project to the interim improvement projects that have led now to the Panama Canal Expansion Program, this book gives a good view of how good and bad or nonexistent project and risk management practices can create project outcomes that are long-lasting and very tangible.

For example, the first Panama Canal project by the French made gross assumptions about canal construction in Panama based on the French experience in completing the Suez Canal in 1869. The project was a commercial venture with many investors. However, the leader of the project was not technically trained, and he failed to scope out the Panama Canal Situation and to define the many "risks" that Panama would present. As a result, the cost estimates were updated almost weekly as the project team learned more and more about what they did not know in the beginning.

The story of the French project is almost the same story as some of the nuclear power plant projects in the United States during the last fifty years. Cost escalation was astronomical and eventually the project was shut down, which is a failure to really define and carry out a realistic project plan for a canal. The failed project had such far-reaching percussions that some have said that it led to a new French government in 1892.

The second project was an American-sponsored project by Teddy Roosevelt and the United States Army Corps of Engineers. Several chief engineers were employed on this project which did a better job of assessing "risks" and scope than did the French project. Nevertheless, the successful American project was a story of new technology development of dam building and electricity systems and controls.

Today's Panama Canal Expansion Program is a $5 billion program with megaproject management and extraordinary risk management techniques being applied. It was interesting to see the program completion in 2014 to ascertain how far project and risk management practices have advanced.

IT IS AS IMPORTANT TO FOCUS ON RISK MANAGEMENT AS IT IS TO FOCUS ON PROJECT LESSONS LEARNED IN YOUR PMO CAPABILITY

In recent discussions with several evolving and maturing PMO organizations, it was obvious to me that they were wrestling with the question "What is more important: To have a robust Project Risk Management Plan in place or an actionable Project Closeout and Lessons Learned Process?"

As you know, I have devoted much time and effort to developing a Project Closeout and Lessons Learned Framework which is aimed at providing an "actionable" process improvement output for an organization's project group. I have also written about the relationship between a project lessons learned process focused on "Significant Events" in projects to an active Risk Management Plan for project work. Talking to many in the project community, I believe that some have concluded that I believe that an "actionable" Project Lessons Learned process is more valuable than a Risk Management Plan.

What I would like for those to understand is that a Project Risk Management Plan is just as important as an actionable Project Closeout and Lessons Learned Process. As I have discussed in the past, most PMOs evolve by developing a "once through" project management methodology that offers the organization specific outcomes or deliverables from its projects. Once PMOs digest the feedback from stakeholders, customers, auditors, and others in the organization who have valuable input into a PMO's competencies, capabilities, and development, the PMO begins to introduce additional capabilities that justify themselves on a cost-benefit basis. Such capabilities as vendor management and risk management are typically introduced in the methodology to provide additional benefits to the project's successes.

This is typical of the way maturing organizations operate. As they recognize the need to build in capabilities, they incorporate those capabilities in the next "build out" of their facilities like the project presentation rooms. Likewise, in maturing a PMO building in risk management is a given so that it can contribute positively to every project.

Risk management and project lessons learned are two competencies and capabilities that need to be in place for a really successful project outcome and success.

HOW DOES AN ACTIVE PROJECT RISK MANAGEMENT PLAN FACILITATE AN "ACTIONABLE" PROJECT LESSONS LEARNED FRAMEWORK?

In the course of my work developing an "actionable" framework for Project Lessons Learned, I have learned that one of the great facilitating actions contributing to a robust and "actionable" Project Lessons Learned Framework is an active **Project Risk Management Plan**.

Many project organizations already incorporate risk management as a capability in their planning and execution of projects, but many others have not reached the stage where they really appreciate how a Risk Management Plan can contribute to project control, outcomes, and success.

Although it is not necessary to have a Risk Management Plan to apply the Project Lessons Learned Framework I have developed, the real power of having a Risk Management Plan lies in the fact that "Significant Events" that may become "Candidates" for lessons learned may already have been identified in Risk Mitigation Planning and Risks Triggered during a project. That makes it easier to collect as many real significant events as necessary to complete a project lessons learned exercise.

Risk Management is, of course, an attempt to anticipate actions and events in a project for which the organization has not formally planned, but which could impact the project.

Risk Management is also a way to address a project's assumptions that might previously have been considered to be unchanging during the course of the project. We all know from experience in managing projects, however, that initial "assumptions" almost never stay the same during projects. So, risk management is a technique for planning and taking action for assumptions that change during a project.

Risk Management is an exercise in which you identify possible events that might lead to adverse consequences for a project. Mature project organizations or PMOs that embrace risk management techniques are also more likely to embrace change which, as we have observed in other blog posts, correlates highly with embracing project lessons learned.

Risk Management also usually involves some discussion of how risks could introduce gaps in expected-versus-actual results for a project.

Project organizations that do not formally have a Risk Management Plan in place for individual projects should, of course, still pursue project lessons learned exercises at the close of their projects. As the organizations gain more experience in capturing, documenting, and sharing lessons learned, they will also become more aware of systemic risks in their business or project context which, in turn, should make it easier to introduce a Risk Management Framework that addresses the portfolio of projects and individual projects as well.

Ask yourself whether your organization fully utilizes its Risk Management Plans and Frameworks in assisting in project lessons learned exercises. You can play a leadership role in introducing the link between risk management and lessons learned.

WHAT IS THE ROLE OF RISK ANALYSIS AND RISK MANAGEMENT IN THE MODERN PMO?

> The mastery of risk is the foundation of modern life, from insurance to the stock market to engineering, science, and medicine. We cannot see the future, but by calculating probabilities, we can do the next best thing: make intelligent decisions—and take control of our lives—on the basis of scientific forecasts.
>
> (Peter Bernstein, *Against the Gods: The Remarkable Story of Risk*)

> I am used to thinking three or four months in advance about what I must do, and I calculate on the worst. If I take so many precautions, it is because it is my custom to leave nothing to chance.
>
> (Napoleon I, in a conversation with Marshall Murat, March 14, 1808)

The Deepwater Horizon oil spill crisis in the Gulf of Mexico is a compelling case study about risk. There has been much written about how BP and the

United States should have developed a Risk Management Plan as a contingency in case such an event were to occur. It is obvious that many people feel that the principal parties involved in the exploration, drilling, and production had no Risk Management Plan in place.

I suspect that is far from the truth.

Modern project management practices include risk analysis and risk management as key and integral components of any operating and business plan involving assets and resources.

Every operational move in the portfolio of projects has key risk management activities that are focused on the **two key aspects of risk**:

1. **The probability of an event occurring**
2. **The impact of that event if it should occur**

These aspects are then analyzed based on prior information, other incidents, and events, in order to determine a four-quadrant layout of potential risk outcomes. Typically, companies place the most emphasis on the quadrant of high probability of occurrence and high impact when developing their mitigation and contingency plans.

The truth is that most mitigation plans and contingency plans are never revealed to the public by project teams unless an event sparks a response. But it does not mean that the mitigation and contingency plans don't exist.

Even prior to the project being initiated, most larger companies employ a risk analysis in their overall Internal Audit plan in order to identify risks involving the geopolitical climate for projects, the resource risk, the technology risk, and so on. The Audit universe of projects is then subject to scrutiny as to which possible risks present the greatest exposure to the company. In other words, which possible risks represent the greatest adverse implications for revenue and profit generation (and possibly environmental outcomes)?

This analysis, in turn, identifies whether the company should focus additional key resources on areas of high potential for risk events. For example, if software or hardware development or new technology is considered a high-risk component of a project, then the Internal Audit staff usually provides key IT or other technology resources to guide the project team development in that area.

At the planning stage of major projects, risk is analyzed as an integral part of the Project Charter and Business Case development to alert the Authorizing

Leadership Group of potential upsides and downsides of project activities. Indeed, a key and integral part of the decision to pursue the project often focuses on risk.

Throughout the project, risk is monitored closely. Adjustments are made when new risks emerge, and some risks are moderated by project activities. For example, if the project involves the prove-out of new technology during the course of the project, the new technology development is monitored throughout the project so that mitigation plans can be employed at a moment's notice.

As part of the Lessons Learned exercise at the completion of a project, the Risk Management Plan is reviewed for its completeness and accuracy of depiction of the events of the project.

So far my discussion has been somewhat "textbook" in nature following the usual Planning, Execution, Close, and Lessons Learned aspects of any well-defined process.

But, if we examine the reality of the risk management process with actual practice, we will see that there are two aspects of risk analysis and risk management that we have not introduced to this point—the concepts of **interpretation** and **judgment**.

These concepts are concerned with the leadership aspects of projects. The actual application of interpretation and judgment to real-life risk cases can vary all over the board based on the corporate culture and the context and climate within which risk is defined and managed.

The aspects of "good interpretation" and "good judgment" should be emphasized more in our understanding of risk in projects. Noel Tichy and Warren Bennis in their book *Judgment: How Winning Leaders Make Great Calls* emphasize that judgment is the core—the nucleus—of leadership. With good judgment, little else matters. Without it, nothing else matters. **Risk analysis and risk management without good interpretation and good judgment are devoid of the key elements of project leadership**.

In your PMO, do you have a well-defined risk analysis and Risk Management Plan in place for your programs? **How** are the concepts of interpretation and judgment carried out? **Who** carries them out? Top management? Or the engineers and planners at the heart of the project details?

If your overall approach to risk in your PMO needs examination and updating, take this opportunity to boldly propose new approaches and new viewpoints. The value addition from such exercises can be enormous.

LET'S PUT A NEW FOCUS ON RESOURCE RISK IN PROJECTS

So many projects employ third-party contractors and technology development personnel on projects today. It seems like the norm for getting any large project done. Organizations these days infrequently rely on purely in-house talent to dedicate as resources to projects.

I want to focus on "Resource Risk" for all my project management practitioners today because I have run across so many elements and aspects of "Resource Risk" that PMOs don't organically take into account.

There are basically five different aspects of contractor resource risk that I want to focus on so that you can take these into account when you help plan or execute a project.

1. Contractor Skills—Often it is the case that when projects are bid by contractors, they will identify the highest skilled and competent resource personnel they think will win the contract. However, when the job is actually carried out, often the contractor will rely on a mix of highly competent and "just plain support" personnel to really get the job done.
2. Contractor Availability—In project environments where a contractor has worked for a long time and there is familiarity between the firm and the third party, it is almost assumed that the contractor will supply all the resources that he has supplied in previous work. But when the job actually starts, the contractor may say that he is spread too thin with his high-quality resources, or the same resources that he has supplied in the past, so he seeks to substitute new resources with which the firm may be unfamiliar.
3. Contractor Subcontracting—Unless actually specified by the project award of work, many contractors or third parties will subcontract some work to be able to satisfy requirements among a number of firms looking for services in a given geographic area.
4. Contractor Working Conditions—This relates to a risk that I uncovered when I was working with a large organization that was engaged in a huge project where the site conditions were clearly undefined due to previous work in that area many years ago that had been abandoned and was now being revisited. Awarding a project contract in a geographic area where the contractor may encounter unknown site conditions,

or even hazardous conditions, may result in additional cost and delay in cleanup which was not originally envisioned as part of the project. This results in a "Pay Me Now or Pay Me Later" type situation in which either the firm identifies what hazards may be in the site before project initiation or they will surely incur costs later when the contractor arrives on site for work.

5. Intellectual Property Arising from Technology Development—In addition to the use of third parties, there is a great deal of technology development being carried out with new projects today. The question of who owns the Intellectual Property from this technology development is always a point of contention if you do not address it upfront in the project. It can be a risk if not handled properly. Be sure you understand when you launch your project that the alternative solutions could employ new technology and how you want the resource risk and technology development risk handled.

I hope this discussion prompts you to define "risk" appropriately for your projects and to have a mitigation plan in place for all the high probability/high impact risks you identify. I know that many of your organizations have robust project risk management plans in place and that is the best defense against "unknowns" that may be lurking in the woodwork to grab you as you proceed through the project.

TEN THINGS TO DO NOW TO LESSEN STRESS IN YOUR BUSINESS DURING A CRISIS

Crises such as the Covid-19 pandemic cause concern and physical changes to the way business is conducted in the United States. It is important to remember that crises such as these automatically raise stress levels among business owners and their employees. But there are some easy things a business owner can do to lower the stress level in the organization.
Here are ten items you can focus on in the short term:

1. If you or your employees are impacted financially or socially in your living environment by the crisis, seek specialized mentoring from SCORE mentors who have experience in specific areas.

2. If you have background music in your business facilities, arrange for more soothing or calming music.
3. If you or any of your employees have a craft or hobby, encourage them to devote more time in the day to that activity. If you have a creative activity you engage in such as journaling, painting, music, or other, spend some time each day on that activity.
4. If your business engages in a community recreation activity, encourage employees to participate.
5. Encourage employees to share stories of business activity that demonstrate the values of the organization or create learning experiences for each other.
6. As a business owner, be yourself. Know your role depending on the level of crisis. Authenticity in the business environment promotes confidence among employees that you are competent and concerned. If you need a reference, see Bill George's book *Authenticity*.
7. Give your employees feedback on a scenario or situation that went particularly well for the business.
8. Communicate very clearly to all employees what is expected and how the business needs to conform to local directives if they apply.
9. Show empathy in scenarios where employees have special situations such as children who have specific needs.
10. Create special networking connections including government officials who might have updated information at a moment's notice that would impact the business and employees.

Investigate whether special processes such as Business Continuation and Emergency Response disciplines might apply.

8

Project Business Requirements

APPLICATION OF DESIGN THINKING PRINCIPLES TO THE COMPLETE DEFINITION OF PROJECT BUSINESS REQUIREMENTS

"Pay Me Now or Pay Me Later"

Sounds like a hit song or a cry that something may have gone wrong with the relationship at hand. Turns out that "Pay Me Now or Pay Me Later" describes a number of consequences of project scenarios with which we all are familiar. It was an expression I coined one time in response to a PMO manager with whom I worked who wanted to proceed with a contract for services without adequately reviewing the risks in employing that vendor or the services requested. "Expediency" was the name of the game. Forget the requirements, just deliver the solution. Sound familiar?

In this particular scenario, I used a timeline to lay out for that manager the consequences for two scenarios: one in which the vendor did not deliver because the requirements were ill-defined and one in which we took time to evaluate and define the requirements carefully to "frame" the problem adequately. Of course, this story would not be worth telling if the timeline for the redo and rework ended before the timeline for the complete requirements development.

The manager signed the contract anyway despite my drawings. And, sure enough, I refrained from saying "I told you so."

If you survey project teams, project managers, and key project stakeholders throughout the various industries that heavily employ project management as a discipline to add value to their respective organizations or to

DOI: 10.1201/9781003593645-8

translate strategy into action, you will find that one of the principal problems encountered is the complete definition of project business requirements. I have alluded to this problem in other portions of this book. Good business requirements are the key to a successful and compelling business case which, in most PMOs, is the basis for approval, rejection, or deferral of a project.

Among the many remedies for the lack of adequate business requirements definition which are found in project literature is the use of very disciplined approaches for defining business requirements by use of templates, coordinated facilitation of group meetings, and very detailed descriptions of the "as is" and "to be" project environments.

Every year there is an enormous amount of rework, redo, and replanning of projects due to incomplete and inaccurate definition of business requirements by project teams. What are some of the shortcomings of business requirement development that most project managers and PMO practitioners will recognize?

1. Use of a "once through" approach: The "once and done" syndrome is rampant in business requirement development because of the desire to get on to the execution phase which is most often deemed by project teams as the most important phase. Very little iterative approach and retesting of business requirements (and original project ASSUMPTIONS) take place in actual project work.
2. Abdication of business requirement development by the group most affected by the project: Usually a business-driven project will have functional and business representatives and stakeholders work with the IT group to develop business requirements. In many cases, I have seen the business/functional group abdicate its position in the dialogue in favor of IT or another lead group taking over the entire process. This has the effect of the development of business requirements from only one single point of view or perspective and leads to interpretations and judgments which, if not tested appropriately, will result in poorly defined business requirements.
3. Narrowly defined interface requirements between people/functional, functional/technology, and people/technology interfaces: This results in technical feasibility and market viability often being ignored and the people/functional interface being treated with simplistic ideas for solutions.
4. Requirements that are not truly empathic in their derivation: Often the stakeholders' real requirements do not surface in the dialogue because a

singular perspective is applied by the IT group facilitating the dialogue with the business/functional group. A real visceral understanding of the emotional, social, cognitive, and physical needs of the stakeholder is not defined. Empathy requires looking at things from the user's or stakeholder's point of view without interpretation on the part of the dialogue facilitator about what the stakeholder or user really wants.

5. Poor "dialogue sessions" between requirements gathering group and the stakeholders or sponsors of the project resulting in rework and redo to create a business case that truly reflects the business requirements being targeted.

You can certainly extend or expand this list from your own project experiences, and I encourage you to write down your observations and test them with others in your project community who are subject to the same project environment. You will be surprised at the recurring themes, many of which occur because of the similar approaches in building business requirements that are defined within the structure of your PMO organization and culture.

You may recognize that one reason for the rise in interest in other methodologies such as AGILE and JAD is specifically because these methodologies address some of the shortcomings in the previous ways in which business requirements were developed. Utilizing iterative techniques, incorporating more than one point of view, and prototyping have been accepted as very applicable to many project situations.

I would like to propose that we take a new approach to project business requirements definition based on principles of "design thinking." Design thinking is an approach to innovation. In fact, it has often been called "human-centered innovation" because of the focus on physical, emotional, social, and cognitive needs of the stakeholder or client. Design thinking has been used successfully by consumer and industrial product companies as well as service providers to open up new markets, define new products, identify new partnerships, and solidify relationship management for the parties involved. Design Thinking starts with empathy, which means looking at things from the viewpoint of the stakeholder, user, or sponsor.

Tim Brown, the CEO of IDEO, has written extensively about design thinking, and his reference book on the subject is *Change by Design*. A.G. Lafley and Ram Charan have written about design thinking and its associated facilitating framework known as integrative thinking in their book *Game Changer: How You Can Drive Revenue and Profit Growth with Innovation*.

In a WSJ article early in 2010, Estee Lauder CEO Fabrizio Freda, formerly an executive with Procter & Gamble and a leading advocate of "design thinking" in product development at P&G, said "We don't want to just do the products that consumers want. We want to be inspired by consumer desires and surprise them with products that they don't expect." That takes a deep understanding of the cognitive, social, physical, and emotional needs of the clients, consumers, and customers. But it also applies to project business requirements development.

Here are some specific actionable things you can do from a "design thinking" perspective to enhance your business requirements definition:

1. Listen. Employ "empathic communication" to understand the viewpoints of the stakeholders, users, and sponsors. Listen to the "stories" and "anecdotes" that your stakeholders tell each other. Often, they will reveal their anxieties, worries, and fears along with other needs that are never mentioned in their straightforward conversations. Engage them in "storytelling." Recall the account of the SAP project manager from one of my blogs who spoke to everyone who would potentially be affected by the changes introduced by his SAP program. His method of empathic communication elicited many business requirements and business opportunities of a very detailed nature. It also built commitment from the ground up for the program scope, issues, and approach which the project manager wanted to pursue.
2. Look for analogous situations or scenarios to develop deeper insights into the needs of the customers. Tim Brown, CEO of IDEO, one of the pioneering companies in design thinking techniques often talks about some work they did in redesigning the processes for a hospital emergency room. They talked to so many ER experts about what could be done that they decided to seek an analogous situation. So they studied NASCAR pit crews. Notice the similarities: high pressure, high competency requirements, and ability to work as a team and as individuals in a time-constrained, close-quarters environment. Often sharing the analogous scenario with your stakeholders can stir their innovative and creative instincts. I mentioned in a previous post the analogy between the project manager and the movie director. Look at the similarities between developing a movie script, background, character, and actor development and managing a project's requirements.
3. Prototype the business requirements situation as best you can. This will provide additional insights into "real needs."

4. Expand your questioning of stakeholders to other than just "superusers" of the systems or new processes. Tim Brown often tells the story of redesigning cooking utensils for the kitchen. After continually using experienced cooks for requirements gathering, they brought in a group of children and asked them to put the utensils through their paces in some cooking classes. The dexterity and handling requirements were immediately accentuated because the children had very few preconceived ideas about how the utensils were to be used.
5. Read or view some background information on design thinking and "integrative thinking." Lectures by Time Brown of IDEO can be easily found on the internet. Also, some business school curricula are now embracing "design thinking" as a basic tool for business strategy and requirements gathering. See the work of Heather Fraser at DesignWorks which is the model school for design thinking at The Rotman School of Management at The University of Toronto. Dean Roger Martin's work on "integrative thinking" can be found in his book *The Opposable Mind.*

This is just the starting point for opening up a whole new panorama of requirements gathering techniques and validation. Design thinkers use the whole world experience in their designs. Project managers should do the same.

Now, "Pay Me Now or Pay Me Later" should take on a whole new significance to you. Give me some feedback about your most useful "design thinking" techniques and how you employed them.

WHAT CAN PROJECTS LEARN FROM THE LATEST DESIGN AND MANUFACTURING DISCIPLINES?

An article entitled "Case Study: Design in Cost Reduction," which was cited in one of the Group Discussions on LinkedIn, caught my eye the other day. The article's author discussed how Motorola University (MU) was working with engineers, designers, marketers, supply chain stakeholders, and clients, in what they termed Design for Manufacture and Assembly (DFMA).

Basically, MU draws on the "Cause" and "Effect" principle that fewer parts in a design will lead to less material and less labor cost in the finished product. A collaborated and integrated effort in the design process, followed by a manufacturing process that closely follows the design, will result in lower product cost and lower overall manufacturing cost.

The article cited an example in which a top electronic device manufacturer used the DFMA process, enabling it to complete twelve product redesign projects over a four-month period, resulting in savings of $6.8 million.

The key for us, as project managers, is to recognize that the same principles can be applied in project planning and execution.

A Design for Project Execution and Delivery (DFPED) could be developed whereby the "Cause" and "Effect" principles might read as follows:

Lower project cost, with increased quality of final deliverables, can be achieved by collaborated and integrated project planning and execution, which includes the project team, key designers, marketers, end user stakeholders, and process/project engineers.

This is very similar to having a Balanced Scorecard concept for projects. If you design the underlying organizational and human resource performance processes correctly, this will lead to improved overall project processes which will, in turn, lead to better customer satisfaction with the process and the deliverables, and which will ultimately lead to the project's financial success.

Many of you will say "Yes, but. It takes a very mature project management process and organization to fully collaborate and integrate these teams and processes such that they may fully realize this cause and effect result in the final project delivery."

While that may be the case, every project organization has the capability and the desire to do a better job than they are doing today in delivering projects and value to their organizations. That desire should be enough to motivate the project organization to start an evaluation of their own Design for Manufacturing and Assembly process or, as I have defined it here, a DFPED process.

9

Resilience in Project Organizations

BUILD RESILIENCY INTO YOUR PRACTICES AS YOU BUILD CAPABILITY TO PERFORM

Those of you who saw the NFC Championship game between the Green Bay Packers football and the Seattle Seahawks are certainly asking yourselves the question about how a team can appear to play so poorly for 55 minutes of the game and then win the game to go to the Super Bowl in two weeks. The Seattle Seahawks are certainly an excellent team and certainly a well-coached team but needless penalties as well as a sluggish offense gave the Packers opportunities to build a 19-7 lead late into the fourth quarter. The Packers certainly capitalized on every opportunity they saw to build a lead that looked insurmountable with five minutes to play in the regulation game. But Seattle continued to play "their game," insisting on using the strengths in their tool kit that had taken them to the NFC Championship game.

Finally, with great effort and agility, they took the lead 22-19. A last-second field goal by Green Bay sent the game into overtime. After winning the coin toss, Seattle moved down the field to score the winning touchdown with a final score of 28-22.

What happened? What was the key to Seattle's success after floundering for most of the game?

RESILIENCY: the quality of an individual or organization to continue to pursue a MISSION and VISION despite continual and repeated short-term setbacks.

Louis Tice of The Pacific Institute was instrumental in using RESILIENCY in his leadership and potential development courses for individuals and

DOI: 10.1201/9781003593645-9

groups. Resilience is something that can be taught and visualized in everyday life and activity to the point that it becomes a part of your philosophy of life. Winning takes more than just one successful game victory. It takes a development of skills and capabilities to know how to win in the face of extreme adversity, to keep the end goal in mind while playing through setback after setback that can be demoralizing to the person or team who doesn't know how to "win."

Here's another sports-based example: after Ohio State defeated Alabama in the NCAA football semi-final game a few seasons ago, Coach Urban Meyer said he had never coached a more "resilient" team in his many years of coaching football. Ohio State lost several quarterbacks to injury throughout the season and played the Alabama game with a third or fourth-team quarterback. They lost an early game to Virginia Tech at home in Columbus, and their offensive line seemed very immature and not functioning well together. But "resiliency" and keeping their focus on the end result of being a "winner" led them to the Alabama victory and ultimately to the National Championship in NCAA football. RESILIENCY is learned, not acquired by any other method such as osmosis or "just being a team."

As a project manager, you are often called upon to deal with short-term adversity in your project processes and in your daily project work. Stakeholders can be obstinate; sponsors can be belligerent; budget leaders can be intolerant; and project team members and contractors can be single-minded or share a different perspective of how projects should proceed.

As the PROJECT MANAGER you and only you can build the RESILIENCY into your project team and the project environment. ARE YOU UP TO THE TASK? Being a "winner" at the project manager game dictates that you develop the RESILIENCY to face every short-term setback and move forward to the next challenge.

Not every project you manage will require that you act in the last few minutes before the project derails as in the case of the Seattle Seahawks, but it does require that you build in RESILIENCY while building in skills and capability to perform. These are distinctly different animals, and some project managers don't acknowledge the key role that RESILIENCY plays. Many project managers don't know that being a "winner" means more than just winning that budget battle with your manager every time you meet.

Project management is a never-ending adjustment to your current conditions which are changing all the time. RESILIENCY is your best hope to survive and prosper and introduce sustainability to your organization.

AS A PROJECT MANAGEMENT PROFESSIONAL, WHAT CAN NASCAR TEACH ME ABOUT REFRAMING MY VALUE PROPOSITION?

On Sunday, February 6, 2022, twenty-three (23) NASCAR Busch Stock racing cars squared off at Los Angeles Memorial Coliseum in a 150-lap race worth $2 million. If you are attuned to NASCAR, you will recognize many incongruities in this scenario. A 1/4-mile race track for a sport that has never sanctioned official racers on tracks less than at least 1/2 mile in length. In a market where the nearest sanctioned race is in Phoenix, the cars navigated the track in 13.6 seconds, achieving the average speed exceeding about 70 mph. Close quarters at constrained speeds made it hard to maintain control.

The traditional NASCAR announcers who broadcast the Daytona and Charlotte races were present in Los Angeles, including famed driver Jeff Cordon. Los Angeles Dodger Manager Dave Roberts was the Grand Marshal.

NASCAR reframed itself to capture a new market. Are you prepared to reframe your approach to project management which might provide you with a new value proposition? For example, are you dedicated to revisiting your Project Charter at intervals to update Assumptions as you find new issues in your project? Are you prepared to add new project team members during your project who have new competencies which can rejuvenate your project? Are you prepared to slow your pace of project actions to ensure success in the face of adversity? Will you build resilience for the long term?

It is up to you as a leader to see the big picture

WHAT CAN PROJECT MANAGERS LEARN FROM THE 2012 RYDER CUP GOLF CHAMPIONSHIP?

The Ryder Cup Golf Championship is a three-day event that alternates each year between a European golf site and a United States golf site. The teams are chosen by a captain from the best PGA golfers from the United States and Europe each year. In 2012, it was held at the Medinah Country Club in Chicago.

In the 2012 championship, the European team was down 8 to 4 after the first day of competition. But they battled back to retain the Ryder Cup, which they had won the previous year in Europe. Saturday and Sunday rounds and competitions were very intense, and a crowd estimated at 40,000 spectators was on hand each day to cheer on their respective teams.

What can project managers learn from the Ryder Cup Championship?

Even though golf and project management are very different, there are some conclusions that we can draw from the actions and performance of the participants that would give us some good insight for future project endeavors.

First, the selection of each team is very significant to the overall capabilities and performance of the teams. Each captain must choose complementing competencies and capabilities to make sure that all situations are covered.

Second, "resilience" is extremely valuable as a factor for competition. Being down early in the competition did not deter the European team from being confident and aggressive in their approach to the Saturday and Sunday rounds. "Resilience" means the ability to focus on an end goal or objective while absorbing all short-term defeats and continuing to focus on the end goal.

Third, matching capabilities and competencies to the challenge or task at hand is very important. On the last day of the Ryder Cup competition, when the European team still needed to make up much ground, their captain chose his players very judiciously, according to their capability in "singles" play, which was the format of Day Three. His best "singles" players were stacked at the front so the team could get off to the best start possible on Day Three. Likewise with project managers, understanding strengths and capabilities and when to apply them most successfully is a great asset.

Fourth, I have to think that each captain chose his team's participants because of their work ethic and authenticity. No commentator on the Golf Channel would ever make so bold a statement, but that is my statement. I believe work ethic plays a tremendous factor in success, whether you are a golfer or a project manager or a baseball player. One of the reasons I follow Mark Teixeira of the New York Yankees so closely is because I think he exhibits an outstanding work ethic toward his professional sport. And what is authenticity really? WYSIWYG—What You See Is What You Get. Remember that old expression. When a person appears to be the same way and performs the same way whenever you encounter them, I believe it means a great deal to success and repeated outcomes of an action.

So as a project manager, how do you stack up against this list of factors? Review this list occasionally and share it with others. I will bet that you will get very good feedback and agreement with many of these principles.

CONQUERING LIFE'S GREAT CHALLENGES: DEVELOPING AND FACILITATING A NEW PROJECT DISCIPLINE

John C. Maxwell can be a mesmerizing speaker. When he exhorts you to give your best, you give your best. When he beckons you to take on a great challenge, you say "How soon can I start?"

I just listened to his lesson on "Conquering Life's Great Challenges" for the fifth time, and every time I felt like he was speaking directly to me. Maxwell's theme is that everyone is faced with great challenges in their life, and those who successfully conquer these challenges are enormously enriched and better equipped and prepared for even greater challenges that await them later in their lives.

Preparation is a key ingredient. Keeping a positive outlook about the end result is essential. Keeping the drive going when obstacles seem formidable is essential. But you can do it.

Maxwell says that there are SIX great "ADD-VANTAGES" that one gains from conquering a great challenge:

1. **Adds Self Awareness and Understanding of Self**: Conquering a challenge makes you more aware of what your real capabilities are and helps you solidify your thinking about what you have to contribute.
2. **Adds and Builds Confidence**: Conquering a challenge gives a person great confidence to proceed to the next challenge with the knowledge that you have a process for facing and conquering that next challenge.
3. **Adds Personal Growth and Stretch**: Expands your capabilities to handle key issues.
4. **Adds Momentum**: Creates the drive to move forward proactively with other key challenges.
5. **Adds New Territory and Growth into New Activities**.
6. **Adds Great Value to Others, Their Lives, and the World in General**.

Maxwell goes on to say that, before an individual attempts the challenge, an individual's world is filled with questioning, intimidation, fear, and uncertainty about his own ability to meet the challenge. But completing the challenge creates a sense of breakthrough, encouragement, strength, purpose, and value. What a difference completing a challenge can have on the willingness and proactive behavior of an individual.

In my case, my most recent great challenge was to develop and facilitate a new Project Closeout and Lessons Learned training for an organization that specifically requested the training because it desired to fill a void in its current project process. The organization also wanted to instill a culture that valued the discipline. The organization was itself faced with a great challenge in the form of a major super-project that was taking place over a number of years.

While Project Closeout and Lessons Learned is one of my favorite topics, and one I had been preparing to tackle for many years, my work for this organization spurred me to focus in-depth on the area for a sustained period of time, leading to my own personal development in each of the six Add-Vantages mentioned by Maxwell in his lesson.

At first, like many of us when faced with a new project, I faced all of Maxwell's classic roadblocks. I felt fear and uncertainty that I would be able to deliver, but I forced myself to overcome this fear. After I prepared a training manual and a well-choreographed presentation, I still had to travel to the organization and present the final product. As I was leaving, one of my closest friends said "Just think how much you will be contributing to their capability … continuous improvement and lessons learned will be invaluable to this organization in creating overall project and program success."

Once there, I knew that preparation would be the key to overcoming any remaining fear. I surveyed the training facilities carefully, and I was committed to making the training times as compatible with the participants' normal workdays as possible. Since everyone agreed that 7:30 AM to 4:30 PM were their normal working hours, we adopted that time frame.

I disciplined myself to be ready to fire on all cylinders at 7:30 AM each morning, to have a working lunch prepared for the lunch break, and to be ready to answer the participants' questions and to rephrase any concepts necessary to make sure they understood.

As the class days advanced, my confidence was building—I was very assured that my Project Closeout Framework was sound. Our group discussions yielded some great insights.

In my experience during this training session, I gained a great deal of insight into my strengths. My best times for action are between 6:00 AM and noon. So, it is best for me to capitalize on that time frame for my most significant work of the day. I also learned that I have a tremendous network of willing colleagues in the project community who can be counted on at a moment's notice. I also learned that my Framework applies equally well to contract closeout and project closeout. I learned that a simple feedback diagram showing "Process," "Result," "Lessons Learned," and the feedback to continuous improvement of the "Process" can apply generically to just about any process an organization wants to pursue. What a grand revelation and one that would not have been recognized without the assistance of a flip chart and an attentive class that demanded the very best of my efforts.

So Mr. PMO, when you are faced with your next challenge, take heed to John C. Maxwell's words on conquering great challenges. Not only will you prepare yourself to willingly take on those challenges, but you will also add tremendous value to your PMO in your ability to tackle ventures and projects that management dictates. It will also build "resilience"—a leadership trait that is highly valuable in this risky economic and political climate.

Step up to the plate; you have the ability to innovate, to meet any challenge, to conquer any language barrier, to overcome any cultural conflicts, and so on, if you will only focus on the task and keep a positive attitude in the face of any setbacks.

WHAT IS "RESILIENCY" AND HOW CAN PROJECT MANAGERS DEVELOP THE QUALITY IN THEMSELVES?

If you are a project manager, you should be working every day to develop a quality or capability known as "resiliency." Resiliency is the ability to spring back or rebound from emotional setbacks. Resiliency is "mental toughness." In this context, "mental toughness" or tenacity is the ability to lock on to an end result, goal, or project and to successfully handle all the temporary setbacks that might occur without giving up on the goal.

How many times in your projects have you seen repeated setbacks, either in deliverables not being completed on time, or on budget and resources not being available or competent to complete the task at hand, or "dependencies" from other projects in the form of "gives" and "get" not being available when needed to support your project goals. Often, these temporary setbacks occur in bunches, and we have all said at times "My, what an awful day!!"

But in order to focus on the end result and achieve all the value from the project, you must put aside these emotions and feelings and deal with these temporary setbacks.

The project manager who can address these setbacks successfully and still accomplish his end objectives will be sought-after over and over again for future project work.

Oftentimes we hear the word "ambiguity" in today's organization because so many seemingly aligned initiatives often are really oblique to the direction in which the organization is moving, creating "ambiguity" in direction, motivations, goals, and objectives. The project manager who can address this "ambiguity" with its associated setbacks and still achieve the end value from his project is extremely valuable to the organization and to himself.

So, how should a project manager proceed to develop "resiliency" in his work ethic, practices, and day-to-day activities?

Louis Tice is a coach and counselor who helps men and women achieve their potential in whatever field or discipline they have chosen to participate. He and his wife Diane started The Pacific Institute for the purpose of using research about cognitive skills and self-image to enable high achievement in individuals. Louis Tice has coached business professionals and both professional and collegiate athletes in the area of development of "resiliency."

Using affirmation and visualization techniques, he has coached athletes to develop the mental toughness to overcome temporary setbacks in athletics to achieve the end result in their selected sport. Lou instructs, for example, college football running backs to visualize themselves running through the forest at night, when they are headed in a general direction but cannot see any of the trees and limbs in the forest. As the athlete starts to run in his selected direction, he may run into a limb, and it will hurt. It is painful but, instead of sitting down, he keeps running. And when he strikes the next tree, it hurts but he just slides to the side and keeps running in the selected direction. Then when he runs out of the forest, trees stop striking him, and so he has reached the end goal and never wavered from the end result he wanted to achieve.

Another exercise might be to picture yourself in a well and you cannot get out. As you reach your hand upward, something strikes your hand and you have to pull back. You do it over and over again and eventually the striking stops and you come up out of the well.

Now, as a project manager, visualize yourself running through a maze that represents your project plan or schedule. When you reach an impasse or an unexpected setback, just slide to the side and keep going. It may be painful, but you just keep running through the maze. Eventually, after you have overcome all the setbacks, you will realize that your end goal is right in front of you. And you reach out and grasp it!!!

Of course, this has applications in other parts of life as well. Suppose you are actively involved in a job search and your most recent interview for a program manager position went very well. But then you receive a call saying that you were not selected for any further interviews because they have selected others with industry experience which more directly aligns with the program manager job description. It's just a "no" and not a "forever" although, at the time, it may seem like a huge setback. So you slide to the side and keep looking.

Developing resiliency in your project manager career will allow you to handle temporary setbacks with less stress and greater confidence. You can begin the road today toward being a more resilient project manager. It is a quality that is essential in today's marketplace and in future project organizations.

10

Knowledge Management

PATHWAYS TO KNOWLEDGE MANAGEMENT SYSTEMS FOR PROGRAM MANAGEMENT OFFICES (PMOS)

In my work with Program Management Offices (PMOs) for the past few years, I have developed some observations about PMO maturity and the capabilities and competencies that PMOs have incorporated into their overall organizational makeup. Typically, PMOs are formed in organizations to ensure that the project groups can consistently and repeatedly deliver excellence in their project outcomes. Or, in the case of strict IT PMOs, they are usually formed to provide project strategy and execution of IT projects for the organization. They are also often formed at mergers between two large firms to ensure that systems integration and application rationalization are handled in a systematic manner in order to achieve the expected synergies of the merger.

As PMOs mature in their capabilities and delivery, they receive much feedback from the stakeholders that they serve concerning what they did right and in what areas they need to bolster their capabilities. This feedback comes from auditors, from business functional areas that are served by the PMO, or from systems/applications/infrastructure delivery in the case of an IT PMO. How well the PMO meets overall organizational objectives and goals is very important to the future of the PMO and its growth.

In my experience, two capability areas that most PMOs almost always improve are vendor management and risk management. Although these two areas might have been given simple methodology tasks at the outset of the PMO, it is the feedback from stakeholders and the organization, as well as

DOI: 10.1201/9781003593645-10

PMO benchmarking outside the organization, that determines to what extent vendor management and risk management are improved in both competencies and capabilities.

Technology development is a third area that normally receives much PMO attention during growth and maturity because more and more projects have a technology development component that cannot be addressed on a "one-off" project basis for long if the PMO is to ensure consistency and repeatability of delivery.

At some point in the development of every PMO, the topic of Knowledge Management is raised as an additional capability area that must be addressed. This happens in several ways. First, the larger organization in which the PMO resides may already have a Knowledge Management System that captures, defines, classifies, and documents data and information for the larger organization. The second way in which Knowledge Management systems are often introduced in PMOs is as a result of recurring problems in projects that are not detected until they have hindered the delivery in three or four of the projects. Then, PMOs get interested in capturing these problems, their solutions, and what can be done to prevent them in the future. They may start with a simple Project Lessons Learned initiative to get the ball rolling. Lessons learned are one of the principal pathways for data and information to be captured, stored, and analyzed in a Knowledge Management System.

Michael Koenig, in his paper "What Is KM? Knowledge Management Explained," from KM World, May 4, 2012, explains the difficulty that most organizations have in implementing effective lessons learned processes.

> The implementation of a lessons learned system is complex both politically and operationally. Many of the questions surrounding such a system are difficult to answer. Who is to decide what constitutes a worthwhile lesson learned? Are employees free to submit to the system un-vetted? Most successful lessons learned implementations have concluded that such a system needs to be monitored and that there needs to be a vetting and approval mechanism before items are mounted as lessons learned. How long do items stay in the system? Who decides when an item is no longer salient and timely? Most successful lessons learned systems have an active weeding or stratification process. Without a clearly designed process for weeding, the proportion of new and crisp items inevitably declines, the system begins to look stale and usage and utility falls. Deletion, of course, is not necessarily loss and destruction. Using stratification principles, items removed from the foreground can be archived and moved to the background but still made available.

> All these questions need to be carefully thought out and resolved, and the mechanisms designed and put in place before a lessons-learned system is launched. Inattention can easily lead to failure and the tarring of subsequent efforts.

I have witnessed this difficulty in implementing Project Lessons Learned processes in several PMOs. Typically, "Project Lessons Learned" are perceived by the project managers and teams as a means of "assigning blame" for things that went wrong in projects. Politically, the project managers, and even management, shied away from attempts to identify lessons learned. Certainly, they did not want them documented for others to read and remember.

However, there may actually be an alternative pathway to Knowledge Management systems that most organizations have not leveraged to this point. It's called "project risk management." If your organization has a robust project risk management process in place that identifies risks, monitors risks if they are triggered in projects, and then facilitates an effective mitigation program for them, then focusing on project risks is a natural pathway to knowledge management that the PMO should explore.

In a particular organization where the path between lessons learned and knowledge management seems torturous, then the path between risk management and knowledge management may present a clearer pathway. Of course, in a fully mature organization, the legs of the triangle would represent flows in both directions meaning knowledge and risk management would feed and sustain the other as would risk management and lessons learned.

PMOs will continue to seek ways to capture, document, classify, share, and use knowledge management. What I have provided here is food for thought about the origin of the data and information that they may wish to preserve in their KM systems. Some very mature project organizations have employed what is termed as "knowledge-based risk" whereby they define risk manager roles and feed those roles with knowledge so that they make informed decisions for future project efforts. NASA, for example, has some of these groups. You may want to consider such a structure for your PMO.

AS A PROJECT MANAGER, WHAT DO YOU NEED TO KNOW?

As a human being, I would like to know several things in life.

First, I need to know and understand everything around me.

Second, I need to know the relationships between everything. I need to know the linear cause-and-effect relationship, the reaction that results from my taking an action or introducing an initiative, and the relationship between and among all variables.

Third, I need to know the nonlinear relationships as well. If I take an action, what might be the unintended consequence of this action? I know this happens because I observe it in the world today.

Fourth, I need to know how these relationships change with time. What are the "dynamics" of the process?

And then I stop and realize that none of this can happen as I have described it.

First, new knowledge is being created every day so that what I know today is very limited with regard to what is known in the universe tomorrow.

Second, the cause-and-effect relationships in linear thinking are compounded by the complexity of the systems I am trying to understand.

The third and fourth follow as have the first and second, the nonlinear relationships, and the dynamics of processes and variables make it impossible for this scenario to be played out. Other individuals are taking actions based on their own understanding of the variables that introduce new wrinkles.

But as a project manager, isn't this what you are trying to do every day?

You want to know everything about your project: who is involved, how the variables interact, how an action creates a reaction from the other participants, what can be expected to happen because of nonlinearity, and how does this change over time?

The answer is simple.

First, observe keenly what is happening in your world. Question those things you don't understand.

Second, learn as you go and never stop learning.

Third, recognize and document the processes that make up your business. Recognize Best Practices as they arise and promote them to your colleagues. Listen as project teams describe what works and what doesn't work. Lessons learned from successes are just as important as those from failures.

There is a model emerging here:

Observe
Listen
Learn
Apply
Teach

Curate

Learn from your successes as well as from your failures.

Create the world you want to live and work in

Of course, this is a model of how we conduct our lives every day.

What I have described here deviates from what happens every day, however. Emotion and habit dictate so much of what we experience. Emotion drives economics. We allow habit to invade our daily processes if we are not cognizant.

Project management is no different from navigating the world as we observe it every day. Learn by osmosis, trying new and different approaches and listening intently to everything that moves around you.

WHAT'S RESEARCH WITHOUT PROJECT MANAGEMENT?

Millions of people are engaged in research today. Some of them just want to satisfy a curiosity they have about a question or problem. Many are seeking to satisfy a need of society. Many are digging into a new technology solution which will add functionality to new or existing product or system.

Research may often involve looking at old problems from new perspectives.

Albert Einstein said:

> The mere formulation of a problem is far more essential than its solution which may be merely a matter of mathematical or experimental skill. To raise new questions, new possibilities, to regard old problems from a new angle requires creative imagination and makes real advances in science.

RESEARCH is a fundamental and basic need to meet the demands of modern life. Often research is a key to new innovation which drives our society forward and enables our ability to satisfy our collective needs.

Many who have employed RESEARCH for a time have learned that there is a research process that derives from their interests and their approach to the questions at hand. But the basic question they face is how to turn their research strategy into actionable results and solutions. That is where project management comes into play.

To convert a research question or problem to a project with a fixed start and end date and to devote or dedicate resources and effort to accomplish the tasks and activities that lead to tangible results is the "essence" of project management. And don't forget to conduct a Risk Analysis to identify potential events that may doom your project to failure.

Any researcher can develop their project management skills to meet the needs of their RESEARCH. A resource I would highly recommend is a 2016 CRC text *Project Management for Research: A Guide for Graduate Students* by Badiru, Rusnock, and Valencia. I have conducted many research projects in my career and directed current research students in many fields.

RESEARCH will be our guide to the future. Embrace it and all its potential for our society.

WHAT IS THE IMPACT OF PROJECT ENVIRONMENT ON PROJECT LESSONS LEARNED AND KNOWLEDGE MANAGEMENT?

Every project manager has been faced at one time or another with capturing, documenting, and sharing lessons learned from projects. Besides being a PMBOK Best Practice, it is also a recognized activity by most larger project management organizations and communities in modern project work.

The focus of most Project Lessons Learned activities is on the project itself. What did the project team learn about its project behavior and actions that could be captured for future project managers to benefit from in the project community? Very little work has been done on the other major contributing factor in Project Lessons Learned however. This is of course the "project environment." Every project is subject to a project environment created by the organization and the external environment in which people function every day. The neglect of the project environment as a major factor in driving team behavior and influencing outcomes and Project Lessons Learned means that the focus on continuous improvement in the project community has been almost exclusively on the lessons learned from individual completed projects.

However, the potential exists for far-reaching leveraging actions to be taken regarding the project environment and the "structure" of that environment that could benefit all future projects as well as provide meaningful insights into project team behavior. And the resulting implications for knowledge management are just as great. Knowledge focused on the project environment can provide insights into how we design future project communities which are robust, productive, team inspiring, and which lead to greater success for all projects.

This was the subject of my paper presentation at the 3rd Knowledge and Project Management Symposium (KPM) in Tulsa in August 2008. Using some "systemic thinking" principles from Peter Senge and Daniel Kim, I looked at several projects in a corporate environment that were "playing in the same space." In other words, they were subject to the same project environmental "structure," policies, and procedures. The project team behavior in each case was driven by that "structure" or by the policies in place. So the leveraging factor in improving the performance of those project teams and the projects resided in an examination of the project environment.

11

Project Best Practices and Benchmarking

BEST PRACTICES FOR THE PMO AND PROJECT MANAGERS: FINESSE IN ACTION

A Best Practice is a process that an organization or group has utilized a number of times and achieved success with many times in regard to the objectives and outcomes of the process. Best Practices are usually defined within a group by examining the practices of the group, reviewing and monitoring the results of the process, and then recording the outcomes for future use in other applications.

What I described above is, of course, an ideal situation. In my experience working with a number of PMOs and project groups, they often understand the intellectual nature of Best Practices, and they utilize processes time and again that yield good results, but they fall short of actually defining these processes as Best Practices and, therefore, they lose out on leveraging, capitalizing, and finessing Best Practices to their advantage. That is the theme and message here—How can project groups and project managers take full advantage of internal Best Practices?

This is not my definition but that of Harold Kerzner, a recognized project management leader and teacher. Some people may also think of Best Practices from an industry standpoint. There are many recognized practices in project management that companies have employed, and which have been adopted by companies benchmarking their practices. In this case, I am referring to internally developed Best Practices and the utility of those practices within a firm.

DOI: 10.1201/9781003593645-11

I have used the following example of a Best Practice within my own household as a means of defining exactly what I mean by Best Practices and getting every reader on a common footing.

Within my household, my wife and I have divided the laundry duties into washing/drying and folding/distribution. For a typical load of clothes containing several bath towels, wash clothes, and hand towels, along with other white goods used in our household, at the time of folding the laundry and distributing it, I have developed a Best Practice of folding the large bath towels first so that I can see the other smaller pieces more clearly and fold them more quickly. This allows the process to be completed quickly and smoothly. It also affects the final quality of the folded pieces. Therefore, from a Best Practice standpoint, the outcomes are improved schedule and quality of products.

As I have pointed out in other blogs wherein I have described this household Best Practice, one of the reasons this is a Best Practice is because it would not apply to any of my neighbors living on the same street. It is highly unlikely that the typical laundry load for any of these neighbors would be similar to our laundry load. In other words, the "business context" is different.

But this leads me to the finesse in project management which I alluded to in the title of this blog.

Usually, I fold the laundry with the objective of returning the clean clothes to the household inventory for future use. But suppose I have a different objective or a different interface process with this Best Practice. Suppose, for example, we are preparing to travel to a resort for a short vacation. In this case, I might order the results of my Best Practice differently to mesh well with the interface process. I might put the white clothes in different stacks to pack them in individual suitcases for the trip.

The upshot of this discussion is that I can anticipate how this Best Practice process will mesh with any interface process and impact the inputs to the interface process through "planning" accordingly. Now this probably does not come as any great revelation to the reader, but it is finesse of the form that many project managers don't anticipate or plan for in their everyday project management practices.

Let's take an example. Suppose your project management group has developed an Organizational Change Management (OCM) Best Practices process for introducing the changes from the project in a way that impacts its users in a minimum way. In other words, users of the new process developed by the project are fully involved and participatory in the development and implementation of the new process. Now suppose that this OCM

Best Practice process can impact several different organizations within your company. As the project manager, you might anticipate how the various organizations might react to and incorporate the outputs and results of the OCM Best Practice process so that it is most compatible with the cultures of the different groups affected by the project.

As an example of this concept, suppose your OCM Best Practice process has a plan for one group involving super-user training of the systems applications associated with the new process, and for a different group, you have planned for a series of group presentations outlining how the new process differs from the old, and how each person's role in the success of the process is carried out. As the project manager, you have the opportunity to guide the timing and development of the materials for each group. It is similar to my example of stacking the laundry differently for renewing the household inventory versus packing suitcases for a resort trip.

Now many of you may react by saying, "That's just project management. There is no finesse there." But the very fact that you have started with a Best Practice process makes this different. The consistency and repeatability you have ensured by using a Best Practice has instilled a "quality" to the outputs for the interface process that could not be achieved otherwise. The ingredients that went into making it a Best Practice process will ensure that the interface process is of high quality and delivered with finesse.

As the organization matures this Best Practice process, it will get better and better about delivery. However, the project manager must be cognizant of changes in the project environment or business context that would potentially alter the process. Since OCM processes usually employ resources with different skills, competencies, and capabilities, this finesse will give the project manager more latitude in his planning and accommodation of resources at the right time with the right skills to ensure success.

However, some project organizations do not know that they have Best Practices. They may employ a process over and over with success but never acknowledge that it is a Best Practice for their project group. If this is the case, they also lose the leverage of being able to anticipate and enhance the outcomes to feed the inputs to the interface processes.

When I worked for a previous employer, for example, our principal products were natural gas processing plants built at the well heads of oil and gas facilities that removed the impurities and water from natural gas and rendered a product of sufficient quality to feed the gas utility distribution systems. Every processing facility at the well head had issues with the startup or quality of products produced by the plant. So, the project group employed an engineer whose principal job was to troubleshoot every new installation for startup problems and resolve those problems as quickly as possible to minimize the online startup time and schedule. His startup process was successful every time. They continued to employ it. Did they ever acknowledge that this process might be a Best Practice for the group? No.

That is where the finesse in understanding the power of Best Practices can yield great results. However, most project groups do not go that extra mile to understand Best Practices.

Those project organizations who invest the time and effort in leveraging Best Practices will see rewards in their process development.

Try these ideas on one of your organization's Best Practice processes. I think you will agree with me that it is easy to see how "theory" and "materiality of results" can be easily linked and implemented. Thank you!!

HOW TO CAPITALIZE ON THE LIFE CYCLE POSITION OF YOUR PROJECT TO DETERMINE THE BEST APPLICATIONS OF ENERGY AND ACTIVITY LEVELS FOR YOUR PROJECT TEAM

The majority of my career has been spent actively involved in defining project practices and leading project teams to the successful completion of projects. I have learned a great deal through direct experience, feedback from other project managers, and benchmarking my own organization's practices versus those of others.

But I never cease to be amazed at the sensitivities that are inherent in the project process, and the insights that can be gained by listening to the feedback of experienced, in the trenches, project managers.

This post is about one of those rare sensitivities that can really leverage the energy and activity of your project team at critical times in your project.

Most project management training deals with what I term "steady state" operations or projects that have a daily activity that is very predictable and routine. New project managers are often introduced to a life cycle for projects that, in its simplest form, can take on four phases:

- Initiation
- Planning
- Execution
- Close

What project managers can and should do in this typical life cycle to maximize the success of their projects is often the stuff of Project Gurus who write and direct the activities of many followers each day. But, in some recent interactions with some very experienced project managers, and, based upon a specific example from my own experience, I have learned a new lesson, one that deals with assessing exactly what point your project has attained in its life cycle and then realizing that "dynamics" versus "statics" is the name of the game in managing real projects to successful completion.

In my own experience, my wife and I moved from Oklahoma to Arizona. Although we had moved a number of times in the past as a result of corporate job changes, this was a move we initiated on our own to return to Arizona.

When we lived in Oklahoma, we had weekly trash service offered by the city. The expectation was that each Wednesday morning a trash bin would be placed on the street with the trash generated in the household the previous week. We could usually determine how much trash would be in the bin each Wednesday and whether we would need to place a black trash bag by the side of the bin some weeks. This was allowable by city rules. During normal weeks with regular family activity, we could estimate whether we would have one bin or a bin plus an extra bag at the curbside.

But the move to Arizona, with its accompanying myriad of unopened moving boxes, introduced a new wrinkle in trash planning. We still had the same size city trash bin which was picked up once a week. But the fact that we were at a different part of our standard family life cycle introduced uncertainties in the weekly trash pickup scenario. Each week after the move introduced new variables into the equation: how many boxes had we tackled, what was the content of the boxes, what decisions did we make each week regarding keeping versus discarding the materials, and so on.

Trash pickup became more of a "trial and error" scenario. Planning was haphazard and often not effective. We had fallen into a point in the household life cycle that was dependent on our choices about unpacking and dealing with the moved goods.

Now, it may be obvious to you at this point that "projects" can be the same way. Understanding at what point in the "life cycle" your project resides can be helpful to understanding and planning energy and activity levels for key project personnel.

During the early stages of a project, it may be necessary to expend more energy and activity to provide direction to the project. Personnel or HR issues, along with budget concerns, may require you as the project manager to allocate more resources and expend more energy for clear direction.

Your success as a project manager will depend on your ability to see and act on a life cycle position. This is a subtlety and a sensitivity to real project work that most project managers don't see until it hits them in the face.

Expend your energy where it can be leveraged to create the most value for your project. Use this experience to plan your future projects and see how they flourish.

Now go tackle those projects with renewed understanding and insight.

THE INS AND OUTS OF PROJECT LESSONS LEARNED AS A BEST PRACTICE FOR YOUR PMO

Project managers who fully experience the project process learn and acquire truth, knowledge, decision-making skills, and good judgment. There are three primary methods by which project managers may learn these lessons:

First, "reflecting" is the preferred method because it results in the highest value to the project manager. "Reflection" means focusing attention on or studying an event or outcome to understand its origin and root causes as they apply to new project situations.

Second, "imitating" other project managers' documented, shared experiences is the easiest method by which project managers may improve their skills. "Imitation" means to behave in a manner which mirrors the actions or behavior of others.

Third, "repeating" his or her own bad experiences and unplanned or poor outcomes may also result in the project manager developing his or her skills, although this method causes the most pain and, in most cases, creates the least value addition.

These concepts paraphrase Confucius' fifth-century B.C. quote concerning "wisdom" and "lessons learned" which many of you will recognize from previous chapters. They relate concepts of "behavior," "actions," "outcomes," "experiences," "pain," "ease," "value addition," and "knowledge."

Why is it that project managers refuse to accept the reality that it is more painful to keep repeating the same mistakes in their projects, rather than learn and benefit from the experiences of others?

Lou Tice teaches us two principles of personal growth and development:

1. People act in accordance with the truth as they perceive it to be.
2. People move toward and become like that which they think about.

As Lou Tice suggested, project managers who act as if Project Lessons Learned can have no positive impact on their future success act in accordance with their perceived truth that Project Lessons Learned aren't valuable. Similarly, many organizations have been reluctant to require their project managers to take the time required to reflect upon their completed projects and document their Project Lessons Learned, despite the fact that most PMBOK practices suggest that project managers properly close out projects with an after-action review and documentation of Project Lessons Learned.

Today, however, I believe that this reality is beginning to change. Many companies have begun to take Project Lessons Learned more seriously and are interested in closing out projects with documentation preserving the knowledge created by the project and the experiences of project participants.

On the other hand, what do project managers want to do more than anything else when they successfully complete a project? Those of us who have observed this behavior over time can tell you that overwhelmingly project managers want to get on to that next great assignment, that next great challenge, and that next great project. Rarely do they want to pause and reflect upon what they have just accomplished, or what the organization could gain if they documented and shared their project management experiences.

So, what should be the driving force for properly documenting and sharing Project Lessons Learned?

We all know that most organizations now recognize that there are certain Best Practices—in both their project management processes and their business context—which they employ over and over again. This is to be expected; when an organization experiences a successful outcome using a key Best Practice, the organization is likely to have successful outcomes in the future if it employs that same Best Practice again. Often these Best Practices are specific to that organization's culture, and they fit into the project process naturally in the course of executing projects. Indeed, many organizations are employing Best Practices now intuitively. Few companies, however, are adept at recognizing and employing their own Best Practices.

Just like Best Practices have become—no pun intended—Best Practices within many companies, shouldn't PMOs look upon project lessons learned as having the same potential to lead to "success" in their project work?

Here is a process and "framework" for looking at Project Lessons Learned that will allow the Project Lessons Learned process to become a Best Practice in your PMO:

What would constitute a "capability-based system" for capturing and sharing Project Lessons Learned?

1. There must be some process or mechanism for sorting out the FACTS in stories, experiences, and anecdotes versus the ASSUMPTIONS.
2. There must be a recognized "review" process to identify candidates for Project Lessons Learned.
3. There must be a willingness on the part of project managers and project team members to speak directly, concisely, and with conviction about project events and lessons. This involves a risk-taking attitude that only comes from developing an internal capability in the organization to acknowledge that Project Lessons Learned add lasting value.
4. There must be a "review" process which addresses the following questions:
   - What were the expected results from the action or behavior of the project team?
   - What were the actual results from the action or behavior of the project team?
   - What is the gap between actual and expected?
   - What are the lessons learned to be captured, shared, and documented?
5. There must be an internal knowledge management system devoted to storing Project Lessons Learned documentation so that project managers may easily retrieve and apply the lessons contained therein to new projects.
6. There must be a single person who is the coordinator or caretaker of the Project Lessons Learned process and knowledge management system so that he or she can analyze the documented lessons learned in order to identify any broader lessons learned which may be applied to the policies, processes, and procedures governing the organization's project management processes.

Once you have mastered these basic elements and gained some experience in applying the process to a number of projects, you can begin to add some sensitivities. For example, you can relate Project Lessons Learned to the risk of developing a new technology concurrent with the project within which the new technology is being applied. At the outset of the project, you can establish a plan to prove-out the technology as the project progresses. A lesson learned can then be documented in terms of the risk of the new technology being proved-out successfully during the project. Such a scheme could introduce concepts such as "controllable" and "uncontrollable" risk. "Controllable risk" could be associated with those portions of the technology prove-out where there is a high probability of success.

Likewise, you can look at Project Lessons Learned at the end of each major phase of your project and apply some integrative thinking principles. This allows a reexamination of original assumptions for the project and sets the tone for good project planning for future project phases.

Where does your PMO stand on closing out projects?

BENCHMARKING TO IMPROVE THE PERFORMANCE OF YOUR PMO

I have discussed in this book how individual project and program managers can improve their capabilities and performance through some simple, yet very tangible and actionable, everyday tasks in the project process.

I want to turn now to "benchmarking"—an area that can be applied to improve performance on several scales, whether it be the overall performance of a PMO group, of an individual group within a PMO, or a single person's performance.

Benchmarking is the comparison of a known state or condition of a group or process to another defined standard group or process which, in the eyes of the benchmarker, represents a level of performance or activity that is desired in one's own group, process, or person.

Benchmarking also introduces innovation into your organization.

Now what does this really mean?

In order to have a meaningful "benchmarking" exercise, you must first define an "as is" base case that describes the current situation with regard to a process, group, or person, and which adequately describes the level of performance.

Next, you must seek another reference group which, in your eyes, represents a "to be" state of performance which you seek to emulate within your own group or process.

Where can you locate such a reference group? Outside organizations such as Gartner, Forrester, or the Corporate Executive Board can provide a good starting point in terms of processes, the expected behavior and activities of these processes, and ways in which you could measure the performance of these processes.

If you are examining actual PMO processes, look at the standard-bearer companies like American Express, Marriott, and Eli Lilly & Company for your reference group benchmark. Similarly, if you want to benchmark in other processes, such as Purchasing and Supply Chain Management, look at the titans–Motorola, HP, and Deere & Co.

You can readily see the power of this approach; it can be applied at the total PMO organization level, an individual internal PMO group level, or even at the individual person contributor level within the PMO.

Benchmarking has been successfully employed by thousands of organizations, even when they had little benchmarking experience when they began the process. The key to a successful benchmarking exercise is a desire to improve. Once an individual or a group decides to get better at whatever the process or task of the group might be, the wheels of successful benchmarking are put in place.

You can be a leader in the benchmarking effort of your PMO. It takes a desire to contribute to the improved performance of the group and a capability to perform. As we have seen in previous blog posts, capability is enhanced with education and training. You should seek out knowledgeable references on benchmarking and other companies in the PMO field who exhibit good PMO performance characteristics. Talk to your local PMI Chapter or to a local PMOSIG group. They can assist in your selection of appropriate PMO performance measures.

12

Project Change Management

CHANGE CREATES OPPORTUNITY: IS YOUR PMO READY?

I once worked with a college professor whose reference books on his desk had notations in the margins which consisted of columns of scribbled, handwritten dates such as "8/25/1982." When I questioned him about the notations, some of which consisted of five or six dates in a column in the margin, he said that many of the topics and subjects were really recurring themes in his research work. The only difference was the social or environmental or technical context at the particular time he reexamined the topic. So to remind himself of those "contexts," he placed dates in the columns as documentation of the different perspectives he had experienced.

When I thought back on this situation, it occurred to me that a topic I had written about during my days in the ConocoPhillips Program Management Office (PMO), and which was published in the company's newsletter several years ago, had even more significance to the project community now than it did when I originally wrote it. So, I decided to resurrect this topic, which roughly can be termed "Change Creates Opportunity."

You see, in the turbulent times in which we find ourselves today, with uncertainty at every turn, and no assurance that a well-intended decision-based action will, in fact, create the significant long-lasting result that were intended at the outset or at the decision point, we need a theme around which to rally our efforts.

I have worked for five major petroleum companies over a twenty-five-year period, and I was involved in at least three major mergers. Accordingly, I often have thought that my middle name is "transition."

DOI: 10.1201/9781003593645-12

Here are some resources that I used to adapt to the changed, post-merger environment during those transitions.

These concepts were introduced by William Bridges in his book *JobShift: How to Prosper in a Workplace without Jobs*. You may also be familiar with Bridges as the author of *Transitions: Making Sense of Life's Changes*, a highly acclaimed book that discusses how people respond physically, emotionally, and intellectually to the transitions and changes in their lives.

The approach set out by Bridges takes advantage of your unique skill set and competencies; tools that you can use to contribute to your PMO's success.

Bridges' main theme is that "change creates opportunity." While change does tend to destroy old opportunities, the bigger implication for you, personally, is that change creates new "needs" within an organization. Many of these needs go unmet until someone recognizes them and takes action. Change relocates the opportunity by changing internal customer needs and the terms under which success is possible.

What does this mean?

An employee should consider the workplace as a "market" with supply and demand forces at work. By looking at the "supply of" and "demand for" services and needs on a continual basis, you can determine what roles and skills are necessary and act accordingly.

In this type of market environment, you must learn to find the needs that are not being effectively or economically met by others—both inside and outside the organizational boundary (because external contractors and vendors can and often do meet these very same needs). This shift in thinking replaces the idea that job roles are restricted to only their formal definitions and allows an employee more autonomy and ownership of their contributions to the PMO.

Change also creates new interfaces. These interfaces may be a face-off between two organizations, between a business and its environment, between two patterns of experience and expectations, or between new technologies and users of old technologies.

Because interfaces juxtapose value systems, assumptions, needs, and languages, they create "unmet" needs. They demand workers who are good at brokering, translating, interpreting, training, linking, facilitating, negotiating, and servicing. These activities bridge the gap in comprehension and familiarity created by the interface. For example, your group may need assistance from the Global Internal Audit group to assess standards and processes in your work area or your project, but no one may be assigned to provide this linkage. The first step would be recognizing the need. The second would be assessing what type of skill or competency is required to fill the gap. The third would be effectively closing the gap.

So, what does this imply for your day-to-day activities within the PMO?

Continually scan the environment for these unmet needs that define the marketplace for your group. Be aware of what skill sets and competencies exist in your group to satisfy a variety of these needs. Act as a facilitator or provide the linkage (REMEMBER THE LINCHPIN?) whenever possible to close the supply and demand gap for services.

Remember: "Change Creates Opportunity."

The market will continually change but by developing skills to recognize, adapt to, and fill the market's changed needs you add value for yourself and the PMO.

Is your PMO ready to meet the "unmet" needs in the marketplace? Is it ready to react to a change as large as a corporate takeover?

The rate of change is increasing.

Today, there is a groundswell of needs being satisfied through social media by varied stakeholders (rather than the traditional supply route of institutions). New external requirements such as Smart Grid, or Sustainability, or Carbon Footprint, or the Green Movement are other examples of forces requiring a change response within the PMO organizational framework.

I encourage you personally to think about the concept of "Change Creates Opportunity." It is a perspective that creates win-win solutions.

HOW DO YOU MOTIVATE YOUR PMO TO MAKE THAT STEP CHANGE TO THE NEXT LEVEL OF PERFORMANCE?

Pretend for a few minutes that you are the vice president of program management for your company and that you are taking some time to reflect on where your PMO has been and where you think it is going. It hasn't always been an easy road getting to this point.

There were times when you thought "project management" might even be a dirty phrase because the promise just didn't appear to be there.

And there was that investment in training and development that you authorized that really hit the budget hard when everyone else was cutting back.

And there have been times when expectations for your PMO have fallen a little short. But, of course, being the dynamic, forward-thinking manager that you are, you recognized that it just took a little "systemic thinking" to uncover the reasons why the group was being held back. And once again, after that intervention, the flow returned to the group.

Now it seems that, with the challenges coming from left and right, and organizational change always in the wind, you are questioning what you can do to motivate this group further so that it can make that next step change. You have authorized all the bonus and salary increases that you could, but with the recession really lingering on in your industry, you are concerned that some of those increases might be negated at the corporate level.

A friend suggested to you the other day that you might take a look at the YouTube video by Daniel Pink based on his book *Drive: The Surprising Truth About What Motivates Us*. You have been putting that off, but maybe it's time to take a look. Can't hurt.

You sit back and watch the video … twenty minutes away from the phone and the other interruptions won't hurt anything.

As Daniel Pink begins to unravel his thesis using a whiteboard and marker, the conclusion hits home rather pointedly. When individuals are provided greater monetary incentives to do rudimentary tasks, the old motivation theory seems to bear out, that is, the more incentive offered, the more the person accomplishes the rudimentary task.

But now, in *Drive*, Daniel Pink is making a new argument. Better pay attention here. When the task requires some cognitive activity and the processing of higher-level complex concepts, the old motivator doesn't hold true. Those who are paid more to accomplish complex tasks actually accomplish less?!! What is going on here? That can' t be. You sit up and take more notice now.

Daniel Pink continues. He goes on to say that when people are given more freedom and autonomy to determine which projects they would work on, and when they are allowed to be self-directed in their execution of those projects, they seem to accomplish significant feats. Then he starts to explain that people who are motivated to master a specific skill (such as a musical instrument) actually go out of their way to achieve this Mastery—even if it means making time in their busy lives to accomplish that mastery.

Then he goes on to say that when highly skilled people are questioned about what really motivates them to accomplish something, invariably they say that they feel that their accomplishment gives them a purpose and that they want to contribute to something larger than themselves.

Many of these people are considered the most competent in their field.

Some of these people happen to be Project Managers.

Daniel Pink summarized by stating that motivation can be thought of as consisting of three "buckets":

1. Autonomy and Challenge—"the desire to be self-directed"
2. Mastery—"the urge to get better at stuff"
3. Purpose—"the desire to make a contribution"

In the video, Daniel Pink cites several examples of companies and organizations which have allowed their employees to take one day a week or month in which they work on anything they desire to work on with whomever they desire to work with, as long as they share it with the others in the group. His research has shown that this has resulted in those organizations showing dramatic improvements in their processes, systems, and other value-adding components (which, to that point, have yielded mediocre results).

Suddenly, it hits you—this video isn't just an explanation of some interesting research—it presents some very startling, actionable results. Some very tangible steps that any organization could take. SOME STEPS THAT EVEN YOU COULD AUTHORIZE, MR. BUSINESSMAN.

What if? Your project and program managers are some of the most skilled practitioners in project methodology, scheduling, team building, and Project Lessons Learned that you have ever seen! Could they actually improve their own performance and raise the level of performance of the entire PMO and the company with something so simple as this? Autonomy … Mastery … Purpose.

You ponder it for a minute and then you turn and pick up the phone.

"George, would you assemble all the managers in the conference room as soon as you can arrange it."

"What's the topic? Motivation and our Program Management Office. We are going to unleash the power of our most skilled Project and Program Managers and take this organization to the next level of Program Management."

Are you ready to take the next step?

WHAT DOES BEING A LIFE-LONG LEARNER REALLY MEAN?

Some people are bitten with it at an early age. Some in later life.

Some experience it through an event that happened directly to them. Some experience it by watching someone else impacted by an event.

Some are touched by a revelation about some aspect of life that they either did not understand before or to which they had not previously been exposed.

No matter how it happens to you ... BECOME A LIFE-LONG LEARNER.

Embrace what you know, but acknowledge what you don't know—especially what you know you don't know.

Choose a "philosophy of life" and let your experiences and relationships guide you to a higher-level understanding of what life is all about.

Lou Tice once said, "People act according to the truth as they perceive it to be." Yes, everyone has a different TRUTH they are living by ... one that is based on how and what they perceive about their surroundings and environment.

However, be very wary that you don't exclude other things from your environment that you need to learn as the world evolves on a day-to-day basis. It is so easy to be a life-long learner about those things you want to learn about, but to be oblivious to other things you need to know. This is a mistake.

In your project organization, have you become a life-long learner? Begin today.

PROJECT MANAGEMENT LESSONS LEARNED: A CONTINUOUS PROCESS IMPROVEMENT FRAMEWORK

Project Management Lessons Learned: A Continuous Process Improvement Framework is intended for project managers, PMO professionals, and students of project management who wish to apply performance-based feedback to their process and project improvements. This feedback is the basis for modern process control theory, which is playing a greater and greater role in process design. Readers can apply the Process Feedback Framework, as well as the lesson learned from the book's case studies, to meet challenges that arise now and in the future.

Special focus is given to technology development in projects and sustainability implications of process design. The book emphasizes the rationale for a robust project risk management program which capitalizes on lessons learned. PMO leaders can use the organizational dynamics and systems archetypes applications presented in the book to define project policy standards and procedures to align PMO behaviors with performance objectives.

Index

For Product Safety Concerns and Information please contact our EU representative GPSR@taylorandfrancis.com Taylor & Francis Verlag GmbH, Kaufingerstraße 24, 80331 München, Germany